Design for Six Sigma

A Practical Approach through Innovation

Continuous Improvement Series

Series Editors:
Elizabeth A. Cudney and Tina Kanti Agustiady

PUBLISHED TITLES

Design for Six Sigma: A Practical Approach through Innovation
Elizabeth A. Cudney and Tina Kanti Agustiady

Design for Six Sigma

A Practical Approach through Innovation

Elizabeth A. Cudney
Tina Kanti Agustiady

CRC Press is an imprint of the
Taylor & Francis Group, an **informa** business

CRC Press
Taylor & Francis Group
6000 Broken Sound Parkway NW, Suite 300
Boca Raton, FL 33487-2742

International Standard Book Number-13: 978-1-4987-4255-9 (Hardback)

Library of Congress Cataloging-in-Publication Data

Names: Cudney, Elizabeth A., author. | Agustiady, Tina, author.
Title: Design for Six Sigma : a practical approach through innovation /
authors, Elizabeth A. Cudney and Tina Agustiady.
Description: Boca Raton : Taylor & Francis, CRC Press, 2016. | Series:
Continuous improvement series | Includes bibliographical references.
Identifiers: LCCN 2016005746 | ISBN 9781498742559 (hard copy)
Subjects: LCSH: Six sigma (Quality control standard) | New products--Quality
control. | Industrial design.
Classification: LCC TS156.17.S59 C83 2016 | DDC 658.5/75--dc23
LC record available at https://lccn.loc.gov/2016005746

Visit the Taylor & Francis Web site at
http://www.taylorandfrancis.com

and the CRC Press Web site at
http://www.crcpress.com

This book is dedicated to my husband, Brian, whose love

and support keep me grounded and motivated.

To my handsome, thoughtful, and funny son, Joshua.

To my beautiful, talented, and driven daughter, Caroline.

I love you with all my heart!

Beth Cudney

To my first born child, Arie Agustiady. Your love for books makes

me want to continue writing every step of the way!

To my princess, Meela Agustiady. You encourage me to

be a better woman, mother, and professional!

To my dear husband Andry, your love and support keep me motivated and driven!

Tina Agustiady

Contents

Preface

To deliver products that consistently meet customers' expectations, it is necessary to develop a process of transforming customers' wants into designs that are useful to the customer. Product design is a process that identifies products' purposes and functions and then allocates them to a structural or concrete form. To shorten product development time and save cost, product teams must evaluate and determine the form of that product at the same time as creating methods for achieving those purposes. The ability to evaluate conformance of potential designs to design specifications prior to hardware builds can shorten the development cycle and save cost. The iterative process of design–build–test–fix is simply too slow and costly.

In order to achieve on-target design right the first time, Design for Six Sigma has been developed to complement the product development process. It typically consists of a set of voice of customer interpretation tools and engineering and statistical methods to be used during product development. The main objective of Design for Six Sigma is to "design it right the first time" by identifying product features and functions that the customer can recognize as being beneficial and to ensure that the design can consistently deliver exceptional performance.

Design for Six Sigma is needed for new processes and companies looking to innovate. To redesign an existing process or design a process from the ground up, success is dependent on a rigorous process and methodology. Design for Six Sigma ensures that there are minimal defects in the introduction of new products, processes, or services. This methodology uses the customer's critical-to-quality characteristics for completion and implementation, ensuring there is user satisfaction. To improve customer satisfaction and net income, we must use the methodologies to institute change, make decisions based on analysis, gather data, and ask the appropriate questions. The consequences of not using the Design for Six Sigma methodology properly can include low return on investment and poor innovative solutions. The objective of this book is to explain how the Design for Six Sigma methodology begins with defining the problem or opportunity, which then leads to two paths: (1) if it has never been done before, design it right the first time using Design for Six Sigma and (2) if it already exists and needs to be improved, use reactive tools such as define, measure, analyze, improve, and control to proceed. Design for Six Sigma can be used to understand customer requirements, consider current process capability, optimize performance, and verify predictions to meet or exceed customer expectations.

Authors

Dr. Elizabeth (Beth) Cudney is an associate professor in the Engineering Management and Systems Engineering Department at Missouri University of Science and Technology. Dr. Cudney has a BS in industrial engineering from North Carolina State University, master of engineering in mechanical engineering with a manufacturing specialization and master of business administration from the University of Hartford, and a doctorate in engineering management from the University of Missouri—Rolla. Her doctoral research focused on pattern recognition and developed a methodology for prediction in multivariate analysis. Dr. Cudney's research was recognized with the 2007 American Society of Engineering Management (ASEM) Outstanding Dissertation Award.

Prior to returning to school for her doctorate, she worked in the automotive industry in various roles including Six Sigma Black Belt, quality/process engineer, quality auditor, senior manufacturing engineer, and manufacturing manager. Dr. Cudney is an American Society for Quality (ASQ) certified Six Sigma Black Belt, certified quality engineer, certified manager of quality/operational excellence, certified quality inspector, certified quality improvement associate, certified quality technician, and certified quality process analyst. She is a past president of the Rotary Club of Rolla, Missouri. Dr. Cudney is a member of the Japan Quality Engineering Society (QES), the American Society for Engineering Education (ASEE), the American Society of Engineering Management (ASEM), the American Society of Mechanical Engineers (ASME), ASQ, and the Institute of Industrial Engineers (IIE).

In 2014, Dr. Cudney was elected as an ASEM fellow. In 2013, Dr. Cudney was elected as an ASQ fellow. In 2010, Dr. Cudney was inducted as a member of the International Academy for Quality. In addition, she received the 2007 ASQ Armand V. Feigenbaum Medal. This international award is given annually to one individual "who has displayed outstanding characteristics of leadership, professionalism, and potential in the field of quality and also whose work has been or, will become of distinct benefit to mankind." She also received the 2006 Society of Manufacturing Engineers (SME) Outstanding Young Manufacturing Engineer Award. This international award is given annually to engineers "who have made exceptional contributions and accomplishments in the manufacturing industry."

Dr. Cudney has published over 75 conference papers and more than 50 journal papers. Her first book, entitled *Using Hoshin Kanri to Improve the Value Stream*, was released in March 2009 through Productivity Press, a division of Taylor and Francis. Her second book, entitled *Implementing Lean Six Sigma throughout the Supply Chain: The Comprehensive and Transparent Case Study,* was released in November 2010 through Productivity Press, a division of Taylor and Francis. Her third book, entitled *Design for Six Sigma in Product and Service Development: Applications and Case Studies,* was released in June 2012 through CRC Press. Her fourth book, entitled *Lean Systems: Applications and Case Studies in Manufacturing, Service, and Healthcare,* was released in October 2013 through CRC Press. Her fifth book, entitled *Total Productive Maintenance: Strategies and Implementation Guide,* was released in June 2015 through CRC Press.

Tina Agustiady is a certified Six Sigma Master Black Belt and Continuous Improvement Leader. Tina is currently the president and chief executive officer of Agustiady Lean Six Sigma, which is an accredited organization that provides certifications for Lean Six Sigma programs.

Tina was previously employed at Philips Healthcare as a director, Operations Master Black Belt. Tina drove all continuous improvement projects in the CT/AMI operations function, bringing efficiency and effectiveness to the highest performance levels within Philips Healthcare. She acted as the transformation leader for the two businesses, providing coaching and leadership in the new methodology.

Tina's recent experience was at BASF, serving as a strategic change agent, infusing the use of Lean Six Sigma throughout the organization as a key member of the site leadership team. Tina improves cost, quality, and delivery through her use of Lean and Six Sigma tools while demonstrating the improvements through a simplification process. Tina has led many kaizen, 5s, and root cause analysis events through her career in the health care, food, and chemical industries.

Tina has conducted training and improvement programs using Six Sigma for the baking industry at Dawn Foods. Prior to Dawn Foods, she worked at Nestlé Prepared Foods as a Six Sigma product and process design specialist responsible for driving optimum fit of product design and current manufacturing process capability and reducing total manufacturing cost and consumer complaints.

Tina has a BS in industrial and manufacturing systems engineering from Ohio University. She earned her Black Belt and Master Black Belt certifications at Clemson University.

She is also the IIE Lean Division president and served as a board director and chairman for the IIE annual conferences and Lean Six Sigma conferences. She is an editorial board member for the *International Journal of Six Sigma and Competitive Advantage.*

Tina is an instructor who facilitates and certifies students for Lean and Six Sigma for IIE, Lean Sigma Corporation, Six Sigma Digest, and Simplilearn.

She spends time writing for journals and books while presenting for key conferences. She is also the coauthor of *Statistical Techniques for Project Control and Sustainability: Utilizing Lean Six Sigma Techniques* and *Total Productive Maintenance: Strategies and Implementation Guide* and the author of Communication for Continuous Improvement Projects.

Tina was a featured author in 2014 for CRC Press: http://www.crcpress.com/authors/i7078-tina-agustiady.

Acknowledgments

Our thanks and appreciation go to all of the Design for Six Sigma, Six Sigma, and Lean team members, project champions, and mentors who work so diligently and courageously on continuous improvement projects.

We would also like to give a special thank you to several people at CRC Press/Taylor & Francis Co. for their contributions to the development and production of the book, including Cindy Renee Carelli (senior editor), Jennifer Ahringer (project coordinator), and Cynthia Klivecka (project editor).

1

Design for Six Sigma Overview

Continuous improvement is better than delayed perfection.

Mark Twain

Design for Six Sigma (DFSS) is a roadmap for the development of robust products and services. The DFSS methodology provides a means for collecting and statistically analyzing the voice of the customer (VOC), developing product concepts, experimenting for designing in quality, product modeling to reduce risk through robust design, and data-driven decision-making for continuous improvement in products, service design, and process design.

DFSS enables users to

- Quantitatively and qualitatively identify and effectively communicate a product concept
- Utilize statistical analysis and quality methods to analyze the voice of the customer
- Interpret the results and make recommendations for new product requirements to meet customer requirements
- Design baseline functional performance of a proposed product concept
- Optimize design performance of a proposed product concept
- Quantitatively verify system capability of a proposed product concept
- Document and demonstrate product design meets or exceeds customer expectations

DFSS should be an insight to creative processes by examining critical design elements. DFSS should be a regular part of any design activity. The focus of DFSS is to emphasize the usability, reliability, serviceability, and manufacturability of the design. It is important when utilizing DFSS to document the comprehensive and systematic design criteria created through the process. It is also important to ensure the production of an adequate design that complies with the customer requirements along with business requirements.

Technical design reviews (TDRs), also known as tollgates, are critical during the design process to ensure all problems are escalated appropriately.

During these TDRs, subject-matter experts should be leveraged using open communication that may stem from difficult questions. Being prepared during these open engagements will enable a successful TDR. There should be opportunities to enhance learning and accelerate knowledge levels for all team members working on the design.

The phases and timing for TDR reviews are as follows:

- Concept phase
- Development phase
- Evaluation phase

The concept phase is conducted before starting a full-scale development so that any significant changes will not affect the scheduling. The concept phase ensures that the design concept is appropriate and all technical aspects are understood.

The development phase is conducted after the design is finalized. The development phase ensures that all risks have been mitigated and the new design is ready for implementation.

The evaluation phase is conducted to ensure all of the product's specifications have been adequately tested and reviewed before a pilot implementation.

Documentation should be kept for all TDRs ensuring descriptions and resolution plans. This documentation should be reviewed at each TDR to ensure the process is moving forward in the right direction and all parties are involved and satisfied.

Six Sigma Review

DFSS was built upon the Six Sigma methodology. Six Sigma is a customer-focused continuous improvement strategy and discipline that minimizes defects by reducing variation with the goal of 3.4 defects per million opportunities. Six Sigma has been implemented in product design, production, service, and administrative processes across industries. Six Sigma focuses on reducing process variation using statistical tools and was developed in the mid-1980s at Motorola. The goals of Six Sigma are to develop a world-class culture, develop leaders, and support long-range objectives. This results in a stronger knowledge of products and processes, reduced defects, increased customer satisfaction, and increased communication and teamwork.

Six Sigma is a common set of tools that follow the five-phase approach of Define, Measure, Analyze, Improve, and Control (DMAIC). The purpose of the Define phase is to define the project goals and customer expectations and requirements. The Measure phase is the stage in which the Six Sigma team quantifies the current process performance (baseline). In the Analyze phase, the root cause(s) of the defects are analyzed and determined. The Six Sigma team eliminates or reduces the defects and variation in the Improve phase. Finally, in the Control phase, the process performance is sustained for a given period of time, typically three to six months, to ensure a culture of change and prevent backsliding. Figure 1.1 describes the questions and associated tools for each phase of the DMAIC methodology.

DFSS

DFSS is a data-driven quality strategy for designing products and services and is an integral part of the Six Sigma quality initiative. The goal of DFSS is to avoid manufacturing and service process problems using systems engineering techniques. The purpose of the methodology is to design a product, process, or service right the first time. Design typically accounts for 70% of the cost of the product, which results in considerable resources spent during the product-development process correcting problems. DFSS is used to prevent problems and provide breakthrough solutions to designing new products, processes, and services.

DFSS embeds the underlying principles of Six Sigma in order to design a process capable of achieving 3.4 defects per million opportunities. The focus is on preventing design problems rather than fixing them later when they can impact the customer. DFSS consists of five interconnected phases—Define, Measure, Analyze, Design, Verify (DMADV)—that start and end with the customer:

Define: Define the problem and the opportunity a new product, process, or service represents.

Measure: Measure the process and gather the data associated with the problem as well as the VOC data associated with the opportunity to design a new product, process, or service.

Analyze: Analyze the data to identify relationships between key variables, generate new product concepts, and select a new product architecture from the various alternatives.

Design: Design new detailed product elements and integrate them in order to eliminate the problem and meet the customer requirements.

Validate: Validate the new product, process, or service to ensure customer requirements are met.

Figure 1.2 describes each phase and the associated tools for each phase of the DMADV methodology.

Phase	Questions addressed	Tools used
Define	• Who are the customers? • What are the customer requirements? • What processes are involved? • What is the business case/need/problem statement? • Who is the process owner? • Who are the team members? • What resources are required? • Which process(es) is the highest priority to improve? • What data supports the decision (metric)? • What are the project goals? • What is the scope of the project?	• Project charter • Voice of the customer • Stakeholder analysis • Process flow diagram • Suppliers, input, process output, customers (SIPOC) diagram • Project plan • Responsibilities matrix • Communication plan • Team ground rules
Measure	• How is a defect determined? • How is the process performed? • How is the process performance measured? • How will data be collected? • Is the measurement system accurate and precise? • What are the customer driven specifications for the performance measures? • Is the process capable? • What are the sources of variation in the process? • What sources of variability are controlled and how?	• Process flow diagram • Measurement system analysis • Data collection plan • Benchmarking • Cause and effect diagram • Process capability • Sigma level • Voice of the customer • Operational definitions
Analyze	• What are the performance objectives? • Which process steps are value/non-value adding? • What are the key variables affecting the average and variation of the performance measures? • What are the relationships between the key variables and the process output? • Is there interaction between any of the key variables?	• Histogram • Pareto chart • Run chart • Cause and effect diagram • 5 whys • Value stream map • Regression analysis • Hypothesis testing
Improve	• What are the key variable settings that optimize the performance measures? • At the optimal setting for the key variables, what variability is in the performance measure? • What are potential solutions to improve on target performance and reduce variation? • What are the potential failure modes? • What improvements should be made?	• Brainstorming • Process failure mode and effects analysis (FMEA) • Poke-yoke • Design of experiments
Control	• How will the process be controlled? • What standards and procedures should be implemented? • How much improvement has the process shown? • How should the improvements be handed off to the process team and process owner? • How much time and/or money was saved? • How will the process be monitored long term?	• Statistical process control • Control plan • Cost analysis • Process capability • Sigma level

FIGURE 1.1
Six Sigma phase descriptions.

Phase	Phase description	Tools used
Define	• Define customer (internal and external) • Define customer requirements • Gather needs • Identify the business case for the project • Create project charter • Develop project plan • Form the team • Translate needs to critical to satisfaction (CTS) • Translate CTS's to functional requirements • Assess technology • Develop plan • Assess risk	• Project charter • Responsibilities matrix • Communication plan • Team ground rules • Voice of the customer • Stakeholder analysis • Data collection plan • Survey design • Quality function deployment • Project plan • Kano model • Product technology road map • Balanced scorecard • Measurement systems analysis
Measure	• Translate customer requirements into engineering requirements • Translate functional requirements to design parameters • Develop/evaluate design alternatives • Resolve design conflicts • Flow down system design to subsystems • Design for reliability and maintainability • Mistake proofing • Assess risk	• Benchmarking • Voice of the customer • Operational definitions • Quality function deployment • Pugh concept selection matrix • TRIZ • Design scorecard • Design failure modes and effects analysis • Axiomatic design • Standardization
Analyze	• Develop transfer functions • Develop system capabilities • Generate concepts • Evaluate concepts • Assess design gaps • Select concept • Assess risk	• Brainstorming • Quality function deployment • Design of experiments • TRIZ • Pugh concept selection matrix • Design scorecard • Process verification • Design failure mode and effects analysis • Process failure mode and effects analysis • Reliability testing • Measurement systems analysis
Design	• Optimize the product or process design for robustness • Optimize tolerances • Demonstrate process/product capability • Mistake proof design • Assess risk	• Brainstorming • Design failure mode and effects analysis (FMEA) • Poke-yoke • Design of experiments • Parameter design • Tolerance design • Design for X • Capability analysis
Verify	• Verify product performance • Verify process performance • Monitor system capability • Implement design and process control plans • Develop transition plans	• Statistical process control • Control plan • Cost analysis • Process capability • Sigma level

FIGURE 1.2
Design for Six Sigma phase descriptions.

DFSS puts the focus up front in the design/engineering process. The key focus is ensuring the team understands the customer requirements and their tolerance to performance variation. To do this, it is necessary to bring the appropriate experts together to engineer a robust solution and reduce the impact of variation.

DFSS is used when

- Products or process do not currently exist
- Introducing new products or services
- Multiple fundamentally different versions of the process are in use
- Current improvement efforts are not sufficient to meet customer requirements
- Products or processes have reached their limit and need to be redesigned for further improvement

Initial process capability limits may not conform to or meet customer needs. Therefore, the DFSS interactive design process is needed to realize customer needs, process capability, and product or service functionality. DFSS enables teams to understand the process standard deviation, determine Six Sigma tolerances, or confirm that customer expectations are met.

Comparison of Six Sigma and DFSS

Six Sigma and DFSS both focus on reducing defects toward a goal of 3.4 defects per million opportunities to improve the financial bottom line. In addition, these methodologies have been successfully applied in a wide variety of industries regardless of the size of the company. Both methodologies are also data intensive and rely on advanced statistical analysis.

Six Sigma is used when a product, process, or service already exists and needs to be improved. In other words, it is used when a product or process is in existence and is not meeting customer specifications or not performing adequately. Typically, organizations can achieve a level of approximately 4.5 sigma through Six Sigma process-improvement efforts until it reaches its limit. At that point, organizations struggle with making further improvement, and, in order to make further improvements, they must redesign the product, process, or service to make it more robust. The existing product or process has been improved; however, it still does not meet the level of customer specification or Six Sigma level. In addition, DFSS is used when a new product, process, or service does not exist and needs to be developed.

Six Sigma focuses on manufacturing and assembly, where the problems are easier to see, but more costly to fix. Since Six Sigma reduces variation

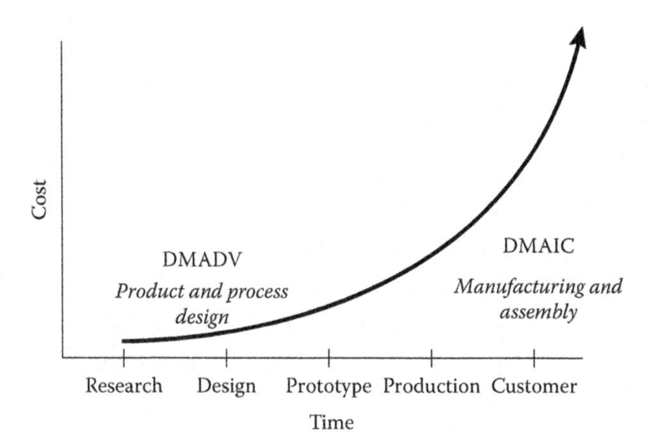

FIGURE 1.3
Design for Six Sigma vs. Six Sigma.

from products or services that are already being produced or offered, it is a reactive strategy. DFSS focuses on product and service design where problems are harder to see and less expensive to correct. DFSS is a proactive strategy since products and services are being designed correctly the first time. Figure 1.3 shows the increase in price to make changes over time as products move from research through design, prototype, and production, and if found by the customer.

The roles in Six Sigma and DFSS are similar. Both methodologies involve project champions, master black belts, black belts, green belts, process owners, and team members. In addition, for either methodology to be successful, there must be full support of upper management. In Six Sigma, these roles can be filled by individuals throughout the organization. However, in DFSS, the role of the project leader, or black belt, is filled by engineering since a new product or service is being developed.

Conclusion

DFSS is a data-driven strategy that focuses on designing a product, process, or service right the first time. It is an integral part of the Six Sigma quality initiative to avoid manufacturing and service problems using systems engineering techniques. The integrated five-phase methodology, Define Measure Analyze Design Verify (DMADV), takes a proactive approach by embedding the underlying principles of Six Sigma to prevent design problems and achieve designs and processes at the target of 3.4 defects per million opportunities. The history and development of DFSS are discussed next in Chapter 2.

Questions

1. What is Six Sigma?
2. When is Six Sigma used?
3. Describe the five phases of Six Sigma.
4. What is Design for Six Sigma?
5. When is Design for Six Sigma used?
6. What department owns or plays the key role in Design for Six Sigma?
7. What are the differences between DMADV and DMAIC?
8. What are the similarities between DMADV and DMAIC?

2

History of Six Sigma

Design is not just what it looks like and feels like. Design is how it works.

Steve Jobs

The development of Six Sigma originated at Motorola in the 1980s. Bill Smith, an engineer, and Bob Galvin, CEO of Motorola, were seeking a way to bring statistics and finances together to improve their operations and processes. With their knowledge and background, they developed a mix of different tools in order to improve the quality of the organization and reduce bottom line costs. The strategy continued to grow and was adopted by Richard Schroeder and Dr. Mikel Harry who decided to transform the process. The thought was to ensure that all companies, especially top operating companies, were utilizing the methodology. Six Sigma is a structured approach to problem–solving that can be used in any business. The concept was to reduce variation while reducing defects in products and services.

There are six main strategies to Six Sigma (Agustiady and Badiru, 2012):

1. Always put the customer first. It is important for a business to know and understand their customer.
2. All management decisions must be based on data-driven facts.
3. Focus attention on management, improvements, and processes.
4. Create a proactive management team that is not firefighting and instead reacting to data.
5. Ensure collaboration within the business especially when decision-making is needed.
6. Aim for perfection.

Six Sigma is best defined as a business process improvement approach that seeks to find and eliminate the causes of defects and errors, reduce cycle times, reduce costs of operations, improve productivity, meet customer expectations, achieve higher asset utilization, and improve return on investment (ROI). Six Sigma aims to produce data-driven results through management support of the initiatives. Six Sigma pertains to sustainability because without the actual data, decisions would be made on trial and error. Sustainable environments require having actual data to support decisions so that appropriate methods are used to make improvements for future generations. The basic methodology of Six Sigma includes a five-step method approach that consists of the following:

Define: Initiate the project, describe the specific problem, identify the project's goals and scope, and define key customers and their critical-to-quality (CTQ) attributes.

Measure: Understand the data and processes with a view to the specifications needed to meet customer requirements, develop and evaluate measurement systems, and measure current process performance.

Analyze: Identify the potential cause of problems, analyze current processes, identify relationships between inputs, processes, and outputs, and carry out data analysis.

Improve: Generate solutions based on root causes and data-driven analysis while implementing effective measures.

Control: Finalize control systems and verify long-term capabilities for sustainable and long-term success.

The goal for Six Sigma is to strive for perfection by reducing variation and meeting customer requirements. Specifications for products, processes, and services should be determined by the customer. Statistically speaking, Six Sigma is a process that produces 3.4 defects per million opportunities. A defect is defined as any event that is outside the customer's specifications. The opportunities are considered any of the total number of chances for a defect to occur. Table 2.1 provides the defects per million opportunities and sigma levels.

A normal distribution underlies the statistical models of Six Sigma as shown in Figure 2.1.

The Greek letter σ (sigma) represents the standard deviation, which is a measure of variation. It marks the distance on the horizontal axis between the mean, μ, and the curve inflection point. The greater the distance from the mean and the curve inflection point, the greater is the spread of values (variation) encountered. Figure 2.1 shows a mean of 0 and a standard

TABLE 2.1

Sigma Level and Corresponding Metrics

Sigma Level	DPMO	Percentage Defective	Percentage Yield	Short-Term C_{pk}	Long-Term C_{pk}
1	691,462	69	31	0.33	−0.17
2	308,538	31	69	0.67	0.17
3	66,807	6.7	93.3	1.00	0.50
4	6,210	0.62	99.38	1.33	0.83
5	233	0.023	99.98	1.67	1.17
6	3.4	0.00034	99.99966	2.00	1.50

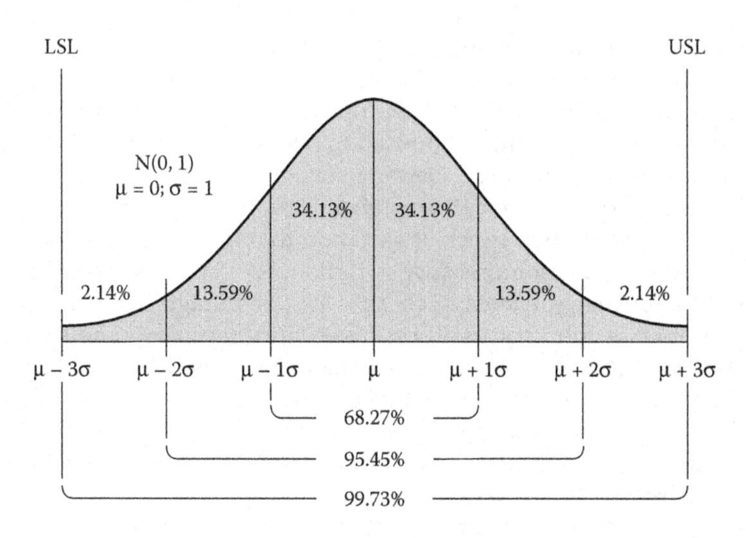

FIGURE 2.1
Areas under the normal curve.

deviation of 1, that is, $\mu = 0$ and $\sigma = 1$. The plot also illustrates the areas under the normal curve within different ranges around the mean. The upper and lower specification limits (USL and LSL) are $\pm 3\sigma$ from the mean or within a six sigma spread. Because of the properties of the normal distribution, values lying as far away as $\pm 6\sigma$ from the mean are rare because most data points (99.73%) are within $\pm 3\sigma$ from the mean except for processes that are out of control.

Six Sigma allows no more than 3.4 defects per million parts manufactured or 3.4 errors per million activities in a service operation. To appreciate the effect of Six Sigma, consider a process that is 99% perfect would allow 10,000 defects per million parts. Six Sigma requires the process to be 99.99966% perfect to produce only 3.4 defects per million, that is $3.4/1{,}000{,}000 = 0.0000034 = 0.00034\%$. Therefore, the area under the normal curve within $\pm 6\sigma$ is 99.99966% with a defect area of 0.00034%.

Variation

Variation is present in all processes, but the goal is to reduce the variation while understanding the root cause of where the variation comes from in the process and why. For Six Sigma to be successful, the processes must be in control statistically and the processes must be improved by reducing the variation. The distribution of the measurements should be analyzed to determine the source(s) of variation and depict the outliers or patterns.

The study of variation began with Dr. W. Edwards Deming, who was also known as the Father of Quality Management. Deming's quality philosophy involved a humanistic approach. He proposed that problems are due to flaws in the design of the system, as opposed to being rooted in the motivation or commitment of the workforce. Deming stated that variation happens naturally, but the purpose is to utilize statistics to show patterns and types of variations. There are two types of variation that are categorized as, special cause variation and common cause variation. Special cause variation refers to out-of-the-ordinary events such as a power outage, whereas common cause variation is inherent in all processes and is typical. The variation is reduced so that the processes are predictable, in statistical control, and have a known process capability. A root cause analysis should be performed on special cause variation to ensure that the occurrence does not happen again. Management is responsible for common cause variation where action plans are given to reduce the variation.

Assessing the location and spread are important factors as well. Location is a measure of whether as the process is centered within the process requirements. Spread is a measure of how much variation exists in the observed values. Stability of the process is required. The process is said to be in statistical control if the distribution of the measurements has the same shape, location, and spread over time. This is the point in time where all special causes of variation are removed and only common cause variation is present.

The *average, a measure of central tendency* of a data set, is the "middle" or "expected" value of the data set. Many different descriptive statistics can be chosen as measurements of the central tendency of the data. These include the arithmetic mean, median, and mode. Other statistical measures such as the standard deviation and the range are called *measures of spread of data.* An average is a single value meant to represent a set of values. The most common measure is the arithmetic mean for normal distributions; however, there are many other measures of central tendency such as the median (used most often when the distribution of the values is skewed by small or large values) or mode.

Special cause variation would be occurrences such as power outages and large mechanical breakdowns. Common cause variation would be occurrences such as electricity being different by a few thousand kilowatts per month. In order to understand variation, graphical analyses should be performed followed by capability analyses.

It is important to understand the variation in the systems so that the best-performing equipment is used. The variation sought after is in turn utilized for sustainability studies. The best-performing equipment should be utilized the most and the least-performing equipment should be brought back to its original state of condition and then upgraded or fixed to be capable.

Design for Six Sigma (DFSS) is a part of the Six Sigma methodology. Six Sigma focuses on the processes to be improved; however, DFSS introduces new products or services. There is an intricate data phase within DFSS that focuses on the generation of ideas to satisfy the customer and ensure innovation. Benchmarking of current processes is imperative for DFSS in order to eliminate unnecessary processes and unneeded steps.

DFSS utilizes a process called DMADV, which stands for define, measure, analyze, design, and verify. The design portion is needed when the current product or service needs to be reinvigorated or a new innovation need is wanted by the customer. The voice of the customer (VOC) is critical for DFSS to be successful in order to understand what the customer is looking for in their business. This concept satisfies the customer, suppliers, and clients, while using creativity to develop the product or service.

The traditional define, measure, analyze, improve, and control (DMAIC) Six Sigma process provides a rigorous process-based approach toward continuous improvement to improve any business process. DFSS is a separate and emerging business process management methodology that has an objective to determine the needs of the customers and the business in order to drive those needs into the product and process solutions.

Conclusion

Six Sigma was developed to provide a structured approach to improve quality and reduce costs. Six Sigma seeks to understand and reduce variation in processes by reducing or eliminating special-cause variation. DFSS extends this philosophy to the design of products, processes, and services. Chapter 3 covers the five-phase DFSS methodology.

Questions

1. What is variation?
2. What are specification limits?
3. What percentage of data is $\pm 3\sigma$ from the mean?
4. What is special cause variation? Give an example.
5. Given the following data set, calculate the mean.

2.98	2.72	3.06	3.04	3.02
2.62	2.71	3.15	3.10	3.06
3.11	2.62	2.90	2.96	2.91
3.16	2.51	3.11	2.98	2.90

6. Calculate the variance for the following population:

37	42	22	27	19
32	39	42	47	31

7. Given the following data set, determine the range of the values.

5.72	4.51	7.28
6.08	3.55	7.83
6.32	8.62	8.17

Reference

Agustiady, T. and Badiru, A. B. (2012). *Sustainability: Utilizing Lean Six Sigma Techniques.* Boca Raton, FL: CRC Press.

3

Design for Six Sigma Methodology

Good design is good business.

Thomas J. Watson

Design for Six Sigma (DFSS) is an emerging business process management methodology with the objective of determining the needs of customers and the business, driving those needs into the product/process solutions. DFSS is relevant to the complex system/product synthesis phase, especially in the context of unprecedented system development. It is process generation in contrast with process improvement.

The primary goal for DFSS is to identify and correctly translate customer needs into proper design choices and critical to quality (CTQ) characteristics; manage significant design risks ensuring that the right design trade-offs are made; and decide the next steps in design. The primary deliverable is to gain approval from the senior managers that the design is solid and ready to move into the next phase of development.

The DFSS methodology will use particular tools that will enhance the value primarily to engineering. The four main priority levers for the DFSS toolkit consist of the following:

- Product design
- Cost/spend
- Value chain
- Marketing

DFSS includes critical thinking skills composed of the following:

- Identify key elements of the design
- Decompose a problem (system) into pieces
- Flow down targets from system to components
- Identify a set of optimal design alternatives
- Prioritize teamwork
- Select between design alternatives
- Manage the risks of a design
- Deliver a robust design
- Manage the variability in the design

- Know where to experiment
- Know when to call for more help (Master Black Belt/technical design review)
- Set and manage project scope and agenda

It is important to identify where the design is most and least sensitive. Figure 3.1 illustrates determining the sensitivities on a measurement system.

The next step is to balance the needs and capability. Figure 3.2 provides an example of balancing the needs and process capability for the crush strength of a product.

DFSS should examine critical design elements, adding peer and expert insight to the creative process. It should also be a regular part of design activity. DFSS should drill down into a particular aspect of a design (usability, reliability, serviceability, manufacturability), software, mechanical and electrical systems, okay since these fall under the comprehensive requirement of a formal design review. More specifically, DFSS with technical design provides a documented, comprehensive, systematic examination of design requirements. Evaluating the adequacy of the design to meet the customer's as well as regulatory/business requirements, ensures a proper examination and the escalation of problems.

DFSS has five tollgates as shown in Figure 3.3.

Optimization is also key to the DFSS process. Each phase is outlined in Figure 3.4 along with the relevant or commonly used tools for each phase. This process is called define, measure, analyze, design, optimize, verify (DMADOV) and consists of the following tollgates and key tasks:

Define

1. Identify customers and their product requirements

Measure

2. CTQ flow down from systems to subsystems and components with specifications
3. Established measurement systems analysis

Analyze

4. Develop conceptual designs
5. Statistically analyze relevant data to assess capability of concepts
6. Develop scorecards
7. Perform risk assessment

Design

8. Develop transfer functions for predictive capability analysis
9. Statistical process control to review variances and process capabilities

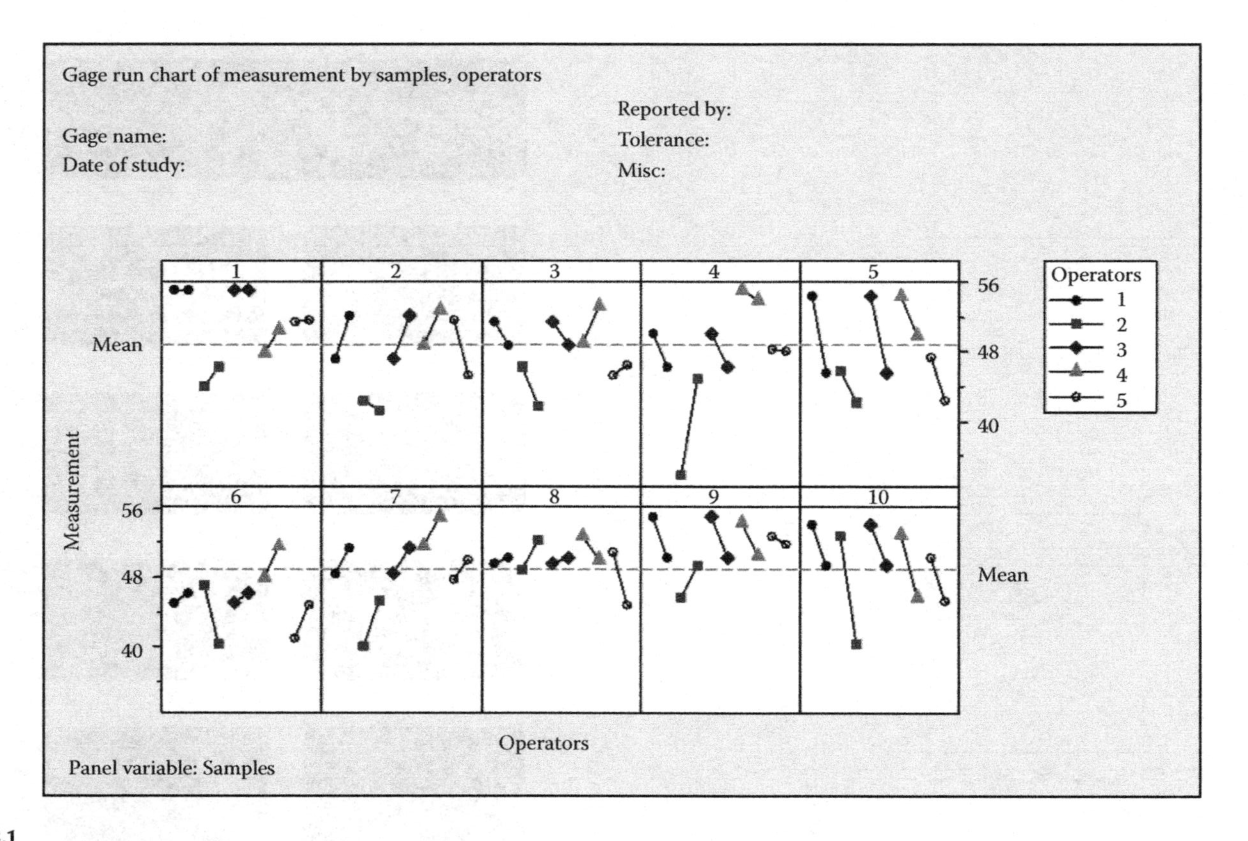

FIGURE 3.1
Design sensitivity.

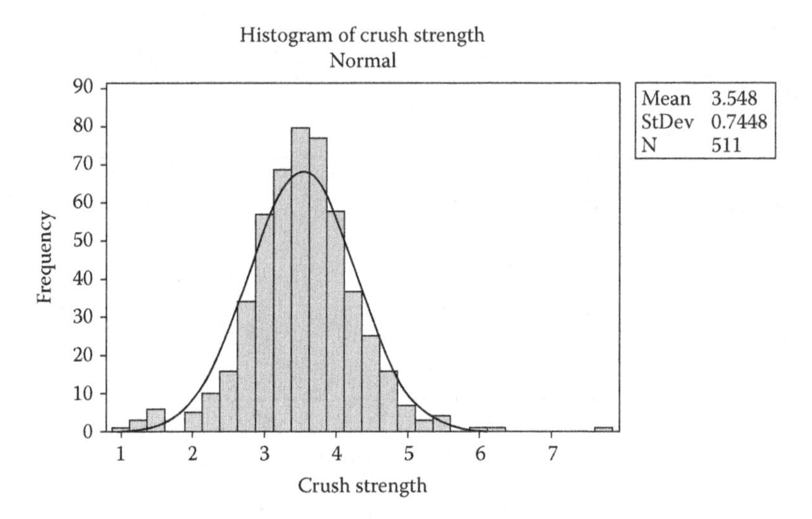

FIGURE 3.2
Determine balancing needs and capability.

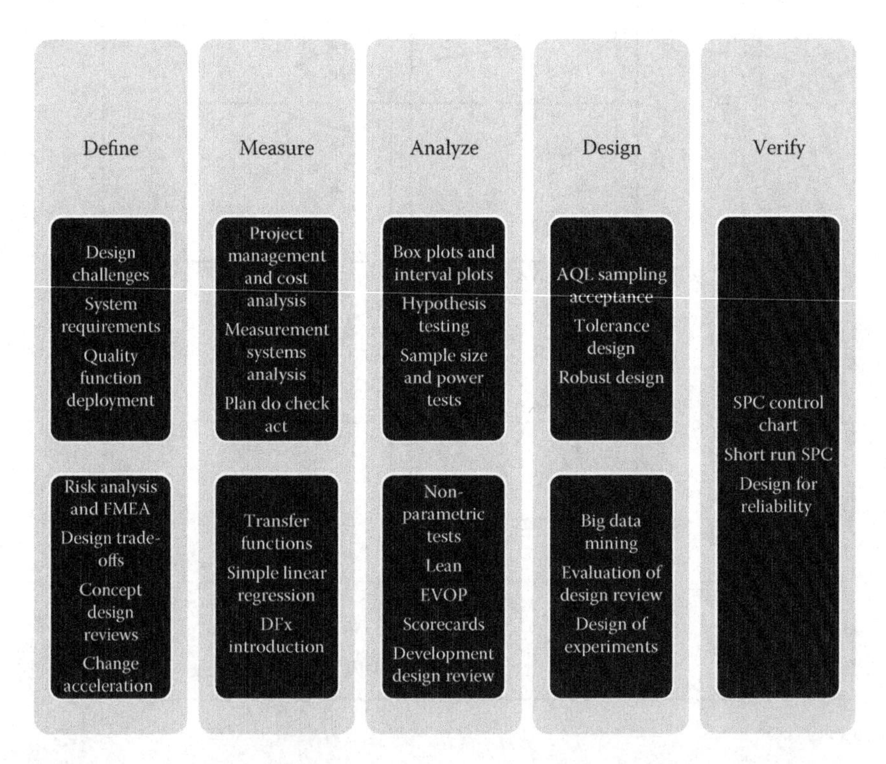

FIGURE 3.3
DFSS tollgates. AQL, acceptable quality level; SPC, statistical process control; EVOP, evolutionary operation.

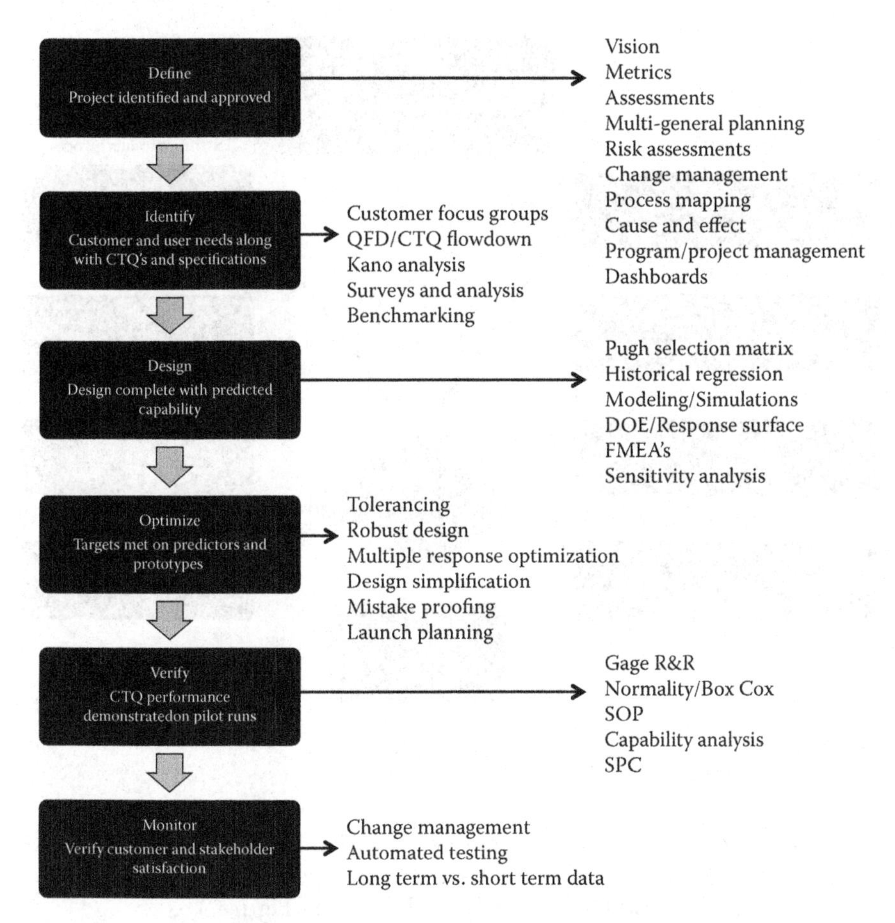

FIGURE 3.4
DFSS phases and associated tools. SOP, standard operating procedure.

Optimize

10. Design to make robust and error proof
11. Tolerance analysis

Verify

12. Statistical confirmation that predictions were met
13. Control plans
14. Documentation, training, and transition

DMAIC and DFSS are two sides of the Six Sigma methodology. They share many of the same tools, as well as an emphasis on data-driven decision-making. The decision point between the two sides of the methodology is symbolic of the fact that while going down one side or the other, we often

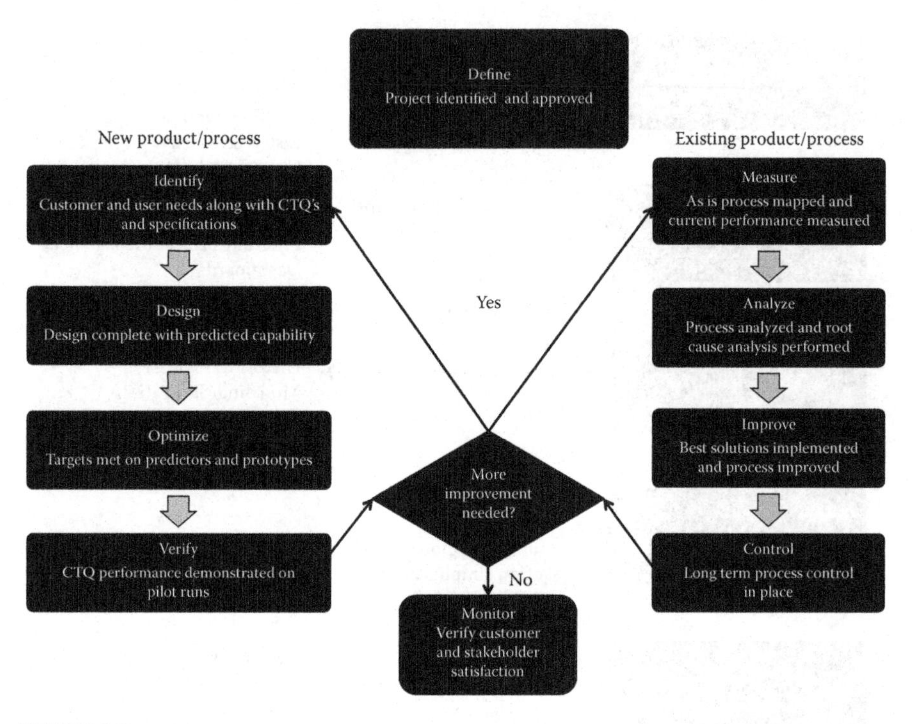

FIGURE 3.5
DMAIC and DFSS relationship.

find that the other side may be more appropriate. It is acceptable to move from one side to the other as the need arises. There are many similarities between Six Sigma and DFSS, which can be seen in Figure 3.5.

Lean and DMAIC both focus on lower cost of nonquality (CoNQ). The focus of Lean is on more efficient production and DMAIC on better quality, so that defects are reduced.

DFSS in contrary, focuses on optimizing the value for the customer. The larger the gap between the created value for the customer and the costs, the better the result.

Lean and DMAIC are a natural fit. They both address processes in terms of optimizing value creation.

Conclusion

The DFSS methodology focuses on determining the needs of customers and the business and then driving those needs into the product/process solutions. DFSS is particularly relevant to complex systems in product development due to its structured, systematic methodology. The five phases of DFSS enable identifying and correctly translating user needs to proper design choices and CTQs; managing significant design risks to ensure that the right design trade-offs are made; and determining the next steps in design. Chapter 4 discusses the culture necessary for DFSS projects to be successful and how to perform a readiness assessment within an organization.

Questions

1. What is the primary deliverable of DFSS?
2. Should everything be at a Six Sigma level of performance in product design? Explain your answer.
3. In DFSS, can executive management take a hands-off approach to the design and structure of the product development process? Explain your answer.

4

Design for Six Sigma Culture and Organizational Readiness

Culture is the habit of being pleased with the best and knowing why.

Henry Van Dyke

Organizational Change Management

Change involves yourself changing along with others. Change can be for the better. However, not changing rarely leads to improvement. Change management relies on the understanding of why things are done and why people are comfortable or uncomfortable with change. Managing others through change processes is needed to change the status quo. This type of change needs guidance, encouragement, empowerment, and support.

What is the status quo anyway? The status quo is defined as the existing state of affairs. In Latin, the meaning is "the state in which." Therefore, maintaining the status quo means to keep things the way they currently are. Some people have the mentality of "If it isn't broken, why fix it?" This methodology rarely leads to change or success, because even if things are going well at the present time, in due time other changes in the world will come into effect that will cause things not to go as well as planned. In a business setting, every corporation strives to be the best it can be in what it does. The competitors then try to beat the best-in-class corporation by doing things differently from and better than their competitor. Eventually, the best-in-class corporation is the one that produces the most satisfying changes, but there must be change in order to satisfy its customers. Keeping things the same way rarely satisfies customers, because they can become complacent and their needs change. Humans desire change and innovation. The desire to be different motivates some to change the status quo. Being complacent normally means being safe and avoiding controversy. Even though this is a safe measure, it will not end with a best-in-class way of doing things, because complacency becomes tiresome, and the thrill of excitement is taken away.

Within Design for Six Sigma (DFSS), the culture of the organization must change for the techniques and tools to be effective. The status quo changes from a mentality of fixing things when they are broken to being more proactive. In addition, DFSS involves everyone within the organization. Therefore, the organization changes from a silo approach based on function to everyone taking an active role in improving products and processes. This change can cause fear in employees. Some employees may feel fear because they sense that they are losing ownership, while others may fear the unknown, since they are being asked to take on new tasks and responsibilities outside their comfort zone.

Effectively implementing change involves frequently incorporating new competencies. New competencies will enable further education and training for all employees. Once this new mentality becomes the norm, changes are viewed as good, and they are always anticipated. The changes will begin with simple items that make work easier and develop into strategic thinking, in which daunting tasks are eliminated by means of more technologically advanced methodologies.

People are afraid of change because it makes comfortable ways uncomfortable while the normal way of operating is changing. This needs to be considered when implementing DFSS. Before DFSS is rolled out, these fears should be addressed by developing a communication plan along with the DFSS rollout. When change is implemented, it should be explained that new skill sets are being taught, which opens doors and increases continuous education. The first reaction of someone being asked to change can be negative, because they are being told to stop doing what they have always done. Ricardo (1995) developed a comprehensive list of why change fails.

- The organization's architecture is not aligned and integrated with a customer-focused business strategy.
- The individual/group is negatively affected.
- Expectations are not clearly communicated.
- Employees perceive more work with fewer opportunities.
- Change requires altering a long-standing habit.
- Relationships harbor unresolved past resentments.
- Employees fear for job security.
- There is no accountability or ownership of results.
- Internal communications are poor.
- There is insufficient resource allocation.
- It was poorly introduced.
- Monitoring and assessment of change progress are inadequate.
- Disempowerment.
- Sabotage.

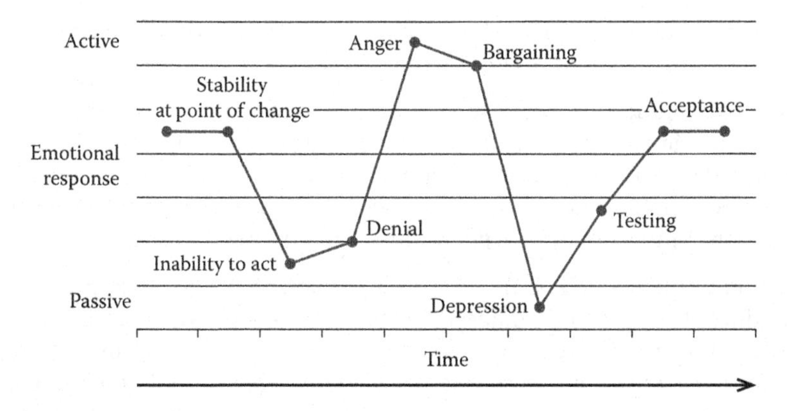

FIGURE 4.1
Change management model.

These are all aspects of fear that should be considered with DFSS. To help people understand why the change is needed, the approach to the DFSS methodology should be explained along with the reason the change is desired. Most of the time, change is needed so that the organization can do things more efficiently while speeding up tasks that need to be completed. Understanding the organization's need for change is imperative in order to embrace the change. This helps the individuals to also understand "what's in it for me?" (WIIFM).

Heller (1999) developed a model for managing change, which is shown in Figure 4.1. The response to change is based on how individuals handle the change. Individuals will go through each type of response, although some individuals may take longer in the different response stages. If a communication plan is developed, and the need for change and how it will affect individuals are effectively communicated, people can move more quickly to acceptance.

In addition, there are several reasons why people resist change, including

- Desire not to lose something of value
- Misunderstanding of the change and its implications
- A belief that the change does not make sense
- Low tolerance for change
- Fear, especially of the unknown that change represents
- Habit and inertia: it is easier to stay the same
- No ownership of the problem
- Not a priority
- Lack of understanding
- Lack of skills needed for the change
- Loss of power

Change requires involved leadership with strong communication skills. The following must be provided while implementing DFSS:

A clear direction and purpose of what is happening, along with focused goals and rationale for the changes being made. People must be willing to let others make mistakes and learn from them without being fearful of blame. Being able to express their feelings openly encourages communication and enables credit to be given for good ideas, good work, and successes.

As you implement DFSS in your organization, you will undoubtedly have individuals who have heard about DFSS, and based on that little information or gossip, they may have heard very positive or very negative things. Spotting resistance to change is important, because negative connotations to change must be eliminated. Spotting change comes from listening to gossip or what people have to say about why things will not work. Observations of people will also reveal whether the employees are resisting change. If employees are being extremely negative, arguing with others no matter what statement is made, missing deadlines or meetings, or not attending or reading up on change assignments, it normally means that they are resistant to change. This resistance must be minimized by addressing the situation quickly. The resistance can only be minimized if the employees have trust in the changes. If the employees are involved in the decision-making and receive consistent and frequent communication, they become less resistant to the change, because they were decision-makers in the changes made, which creates buy-in. Direct, frequent, and consistent communication also makes the changes visible as they happen, with less ambiguity over why the decision was made to implement DFSS. Building positive relationships helps minimize this change by having positive bonds between people working together. The multiple positive attitudes end up being passed on to others with negative attitudes, causing the entire workforce to change. The open exchange of ideas is essential to change. If employees are afraid to bring up their ideas on DFSS, they will continue to resist change, because they are not being heard. It is essential to gain employees' trust and allow them to speak their minds. Once they describe the feelings they are having and the reasoning behind them, it is important to understand the position they are in. By minimizing the fears your employees have about DFSS implementation, you will slowly gain their trust and be able to reassure them that their fears are not going to come true.

In addition, there are several required elements for change:

- Change objectives
- Change vision
- Champion, sponsors, facilitators, teams
- Training
- Systems that enable and support DFSS
- DFSS performance metric(s)

Knowing what is driving the change, the need for DFSS, and how it will affect individuals and the organization is an important step toward acceptance of change. Once the change has been adopted, people will begin to accept the change and be willing to make more changes for the better.

Beckhard and Pritchard (1992) developed several strategies for overcoming resistance, including

- Expect resistance
- Create allies
- Make the case for change
- Choose your opening moves carefully and make them bold
- Articulate the destination; overcommunicate
- Resolve "me" issues
- Involve people to increase commitment
- Use resources effectively
- Promise problems
- Overcommunicate
- Beware of bureaucracy
- Get resistance out in the open
- Give people the know-how needed
- Role model and embody the change values
- Measure results
- Reward change
- Outrun the resisters

Being a role model during change eases the difficulty of the change. People want to be facilitated through change transitions, and need guidance and coaching. To be a role model during the DFSS implementation, it is important to put yourself in the shoes of the other person who needs to complete the change. Thinking about all the reasons why the change can be bad will help with understanding how others are feeling and coping. Giving the business vision of why the change needs to happen will also help guide the scope of the project. Never accept "That's the way it has always been done" as an answer. This answer is an excuse and not a solution. The vision must be clearly defined and provide a direction that everyone understands. It should focus on measurable goals, include the demand for the change, appeal to various learning styles, and answer the "me" questions.

A combination of skill sets will help with challenging the status quo. Different people have different ideas, and combining the ideas to come to a uniform decision will help the group to find strategic opportunities. The group needs to be flexible during the DFSS implementation, and one individual cannot dominate the conversation or the ideas being put forward. Adaptability to other people's viewpoints, schedules, concerns, and styles should be taken into consideration. No one idea is right. Showing other people that you, as a leader, are open to change will help other people to appreciate your values and anticipate the change. Limiting expectations for a perfect workday scenario should be considered in order to accommodate other people and give them the time and attention they need during the time of change. People need a great deal of reassurance during changing times, and that reassurance and commitment from the leader need to be present daily. Leading with values, such that others will respect and follow you, is important, because people do not follow others just because they have the title of manager. The vision needs to be not just stated, but shared and agreed on by all. Leaders need to show that they are not changing the status quo for selfish reasons but for the benefit of the project and the business.

There are several success factors when implementing DFSS. First, it is important to prepare and motivate people. This involves a widespread orientation to DFSS and the necessary training. Management must create a common understanding of the need to implement DFSS. Second, there must be employee involvement. As you implement DFSS, try to push decision-making and DFSS system development down to the lowest levels. As employees are involved with developing DFSS, this will also drive buy-in. Next, it is important to share information and manage expectations throughout the entire DFSS implementation and once it is in use. Finally, it is beneficial to identify and empower champions for DFSS.

Prior to the DFSS implementation, all employees need to understand their new roles. Time lines and targets for implementation should be established and communicated. It is also important to establish easy-to-achieve near-term goals to provide opportunities for success early on.

The timing of change is also important to consider with DFSS. People need time to understand the change and let the new thoughts and ideas sink in. During this time, many questions will arise, and employees should be able to speak with their managers about the questions they have with an open-mind-set mentality. Employees need to be a priority to managers. Without employees, managers are not managers. Competing priorities will lead managers to have little time for their employees. Clarifying expectations of changes and what you, as a manager, are working on will help others understand what they are dealing with. Employees should never feel as though they cannot talk to their managers. Once this fear is instilled in their minds, trust has already been broken, and mending that trust is harder to fix than starting with an open-mind-set mentality. If priorities do change, this should be explained promptly and transparently. Everyone has emotions, so it is important to give the facts about situations rather than react to emotions.

During the DFSS implementation, you should anticipate and celebrate successful results to maintain positivity. This positive feeling can be contagious. The change can be approached from different aspects:

- Positive thoughts approach
- Humoristic approach
- Realistic approach
- Open-minded approach

Maintaining a positive attitude is the most important aspect of changing. Positivity will gain respect and will allow others to see the brighter side of the picture. Humor can ease tensions and bring people together by seeing the lighter side of a situation. While relationships are initially being established, humor helps to show that there is a fun and intriguing aspect to change. Being realistic helps people not to gain false expectations of what could occur, and honesty is well presented. Being open-minded helps people to see all versions of what could happen and appreciate different points of view.

Resistance to Change

All approaches and reactions are different and should be seen with an open mind. Most people resist change rather than embracing it. This is no different when implementing DFSS. This is a problem that must be overcome, and should be seen with the end in mind. Ensuring that people understand the benefits of the change helps to fight resistance. Incorporating different views and ensuring that all perspectives are taken into consideration will help decisions to be made and accepted. Motivating and encouraging others to see that changing the status quo is for the better will help resistance as well. Listening and understanding to all employees will also show that different perspectives are appreciated. Patience is extremely important during implementation. It may take people varying amounts of time to embrace change. Showing patience during the change will help ease the situation and show that there is a realistic approach behind the methodology. It is important to show, share, and ensure that others believe the vision being disseminated. Alignment is key to changing systems and resistance. Creating lists of pros and cons can also show the vision and reasoning behind changing. During the creation of the lists, current, older, and new ideas should also be listed to show why things were done in the past and what benefits DFSS will provide for the organization and the employees. Finally, positive reinforcement is essential throughout resistance to change. Not all people have natural self-confidence. Giving people the confidence they need and coaching them helps build good leaders. Publicizing wins helps people's self-confidence to grow and shows that working through the difficulties will be managed appropriately. Flexibility during this management must be learned by employees, managers, and leaders.

Using Known Leaders to Challenge the Status Quo

Self-leaders can be identified by their ability to influence others and engage in activities. It is to a manager's benefit to use these leaders to help drive change. Assessing the current structure of who the natural leaders are, who the resistance comes from, and who the followers may be will make the process easier to understand and manage. Simple questions can be asked:

- Whom do people listen to?
- Who will adapt to change easily?
- Who will resist change?
- Who will help with the alignment of change?
- Who will motivate others?

Once the support structure is understood, the use of the appropriate people in the proper aspects will align the process of change management.

Identification of problem areas will help leaders to be realistic about what items may not go as planned. Using the known leaders to communicate and distribute the information needed will also help alleviate problems. These people will assist in supporting the structure and processes and maintaining the vision. Goal alignment with known leaders must be established before the leaders are used to motivate others. Once these goals are shared and agreed on, these leaders will dramatically help with change management.

Communicating Change

Communication is the most important aspect of successful changes. Communication prepares others for what is to happen, creates shared and agreed-on visions, and builds relationships within groups. Communicating change should consist of many aspects. Communicating the specifics of what is occurring is important for people to understand and buy in to the change. People need to understand the importance of change to the business and the reasoning behind it. It is important that the people delivering the message themselves understand the reason for the change. The downside of what could happen if the change does not occur should be explained at the same time. Explaining the urgency of the change will speed up progress. Specifics on time lines and milestones also help drive the urgency.

Motivating people to change is imperative in order for employees to feel empowered to change and not just make changes that they will not be able to sustain.

Finally, publicize the wins of work well done or employees who have embraced change. Preparing information and updates on results, changes, and progress will help people visualize the impact the changes are making. It is important to communicate as many of these wins as possible in person. Many people will begin asking questions; admitting that you do not have the answers is acceptable as long as there are means to show that you will find out the information. Being truthful and transparent will help clarify expectations even when discussing the downsides.

Recapping of past communications and topics with their outcomes will help gain credibility for the work that has been performed along with the visible successes.

Many people will want to understand their roles and responsibilities; favoritism should not occur. Consistency is a key for success when dealing with different people and different personalities. Mistakes will happen, so learning from mistakes while being transparent about why they occurred will prevent them from happening in the future. Capturing best practices and learning techniques will help benchmark how processes should work in the future.

Engagement from all levels of employees will help support projects and increase awareness of new initiatives.

Successful transformations and DFSS implementations come from asking the basic questions and blending them appropriately for the proper answers (who, what, where, why, when, and how). How to change and what to change are the key factors that will determine the success of change. Once these measures are established, the implementation process is simplified. Successful workplace changes occur with a proper foundation and framework that have been established by management and employees through a shared vision. People must drive these visions together with their different skill sets. Cross-functional groups working together as teams are what drives successful change. The real change comes from the combined efforts of all employees.

Information sharing during the DFSS implementation is a powerful means of communication. The exchange of information results in the use of the data to make strategic decisions. A company's culture is based on this information sharing. Culture is based on beliefs and values, and the combination and agreement between the two. Wanting to increase productivity and efficiency is part of the culture. If all people believe in this, the vision is the same. If some people have not bought in to the vision or need to change, then a culture shift needs to take place. Once the culture has the same vision, all employees can work toward the same goals.

According to The Steve Roesler Group and Heller (2009), there are five clear-cut messages that show it is time for a change. The following five qualities indicate a need for change:

- *People whom you trust strongly believe you should make a change*: If multiple people whom you are close to think it is time for things to be different, it may be time to listen to them. They may understand you and see that the changes may actually benefit you.
- *You are holding on to something and just cannot let go*: If a particular situation is continually on your mind and you simply cannot let it go, this is seen as a signal. If this is bothersome for you, a change could ease your mind.
- *You feel envious of what other people have achieved*: Jealousy is an evil beast, but it may be able to help you better yourself. If you are envious of someone or something, instead of being envious, you are able to change your ways to become more successful instead of simply being jealous. Taking action toward bettering yourself reduces jealousy and makes you proud of yourself.
- *You deny any problem and are angry in the process*: Anger is seen as a symptom of denial. Looking for help from someone or being able to help someone in need reduces the anger. Increasing communication will help mitigate the problems by allowing individuals to admit that there is a problem at hand. Having an open mind to the problems will encourage you to make a change for the better.
- *If you do absolutely nothing, the problem will continue*: Without addressing the situation or holding people accountable, no change will be made. This is because it is not known that the problem is bothering you, or no one wants to address the problem at hand. The situation must be addressed for change to occur, but this must be done in a professional and calm manner, or it will be seen as negative. Being honest and up front will help others see what changes can be made and will help you to change for the better with the communication received.

There are several simple methodologies for tackling workplace status quo:

- *Communicate where wasted time is being spent*: This involves giving facts on costs and data behind efficiencies. Visual management of key performance indicators (KPIs) will provide the targets and goals and current status.
- *Address costs from doing work repetitively*: Discuss extra labor hours and extra materials spent due to mistakes and having to correct these mistakes.
- *Communicate the need for efficiency improvements*: Companies and people generally need to be more efficient. Communicating the need to increase efficiency because capacity restraints are present is an easy way to show that the goals of efficiency improvement are needed.

Showing competitors' efficiency numbers or predictions will help communicate the baseline of where the efficiency should be.
- *Show that change is easier than people think*: Nobody likes to have a "flavor of the month" of how to do or address something. People must see change as structured and positive. It is important to show that change is innovative and teaches others creative measures. Continuous education is imperative to growth, and changing the status quo will invigorate these behaviors by having an intelligent workforce that is up to date on technology and teachings.
- *Show that change is a habit*: Change should be a daily habit and not just a way of doing something at a particular time. Incorporate change into people's everyday lives so that change becomes something easy and normal. Show that changes are good for personal and professional growth and must be incorporated into both their work life and their personal life.

Dealing with change is simple if people trust why changes are occurring. For the change to take place, it is necessary to explain that change is needed to be competitive and expand while being innovative. Management's beliefs and values need to be shared as a combined vision rather than a top-down approach, so that each individual is on the same page about change. There are four main changes that occur in the workplace:

- New products or services
- Organizational changes
- New management
- New technology

Explaining the need for the different changes and the rationale behind the decisions made will help gain buy-in for the change. Sometimes, the change does not seem to be for the better (e.g., layoffs). It is important when communicating change, even change that is not desired, that a positive outlook is possible. An effective way of communicating any type of change is through factual information and the best information available. Most of the time, the changes taking place are positive and affect cost, processes, and culture.

Cost changes should be presented as being necessary to be competitive in the market place. If cost changes do not occur, market demand will turn to cheaper goods, especially if there is a strong competitor with high-quality products or services. Process changes must be made to achieve continuous improvement. Increasing efficiencies will increase market share by being competitive as well as reducing lead time. Innovation should be sought to keep up with the latest technologies. Cultural changes are normally the most difficult, because these change the way that normal operating procedures take place. Explain that cultural changes are needed, so that there is a common vision for the workforce and no divide between management and other employees.

Implementing successful change involves using the proper resources, whether those resources are things or people. The plan then needs to be established and agreed on so that there is a rationale behind the changes being made. The changes then need to be implemented. Often, people think of great improvement ideas but do not fully implement them, making them impossible to sustain. Finally, communication is the key to change management. All the steps of the process must be communicated to all people and must be agreed on in order to be successful. Effective communication involves communicating often; giving reasons behind implementing DFSS; explaining who was involved in the changes; and giving status updates on the changes that have occurred, what will occur, and the effect of the changes on the business.

Conclusion

As a rule of thumb, plan for setbacks to occur and have a risk management plan. Risks can be threats or opportunities. Brainstorming of potential risks is a key way to begin the problem-solving process. Risk actions can consist of avoiding risks, mitigating risks, acknowledging risks, or accepting risks. There are many tools and techniques that can help mitigate risks, including decision trees, Strengths, Weaknesses, Opportunities and Threats (SWOT) analysis, risk grids, cause and effect (C&E) diagrams, and Program Evaluation and Review Technique (PERT) analysis.

Decision trees use a flowchart structure to illustrate decisions and their possible consequences (Figure 4.2). These diagrams are drawn from left to right, using decision rules in which the outcome and the conditions form a conjunction using an "if" clause. Generally, the rule has the following form:

if condition1 *and* condition2 *and* condition3 *then* outcome

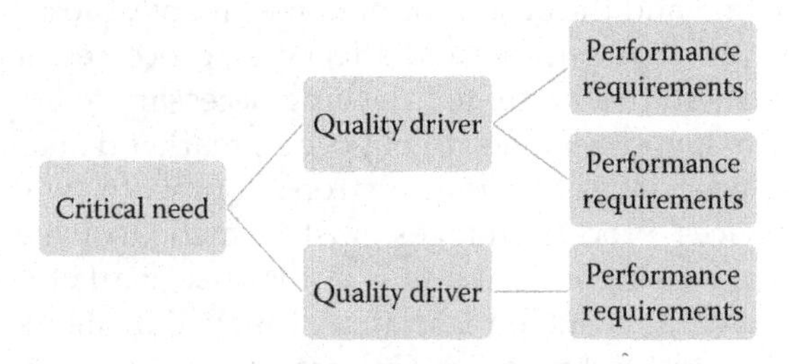

FIGURE 4.2
Tree diagram.

Strengths What do you do well? What unique resources can you draw on? What do others see as your strengths?	**Weaknesses** What could you improve? Where do you have fewer resources than others? What are others likely to see as weaknesses?
Intelligent resources—employment Good intentions to want to do right things Communication technologies Communication philosophies—methodology Organizational direction Strategic executives	Competitive salaries Managers People believe their executives are not aligned—organizational direction Leadership competency Learning gap
Opportunities What opportunities are open to you? What trends could you take advantage of? How can you turn your strengths into opportunities?	**Threats** What threats could harm you? What is your competition doing? What threats do your weaknesses expose you to?
Learning technologies to keep up with the times Using resources properly Generational shifts Types of communication Reverse mentoring New revived energy	Keeping up with technology/new learnings Large generational gaps Creating work environments to engage all generations and keep them attracted Retention

FIGURE 4.3
SWOT analysis example.

A SWOT analysis is a structured approach to evaluating a product, process, or service in terms of internal and external factors that can be favorable or unfavorable to their success. Strengths and opportunities (SO) ask: "How can you use your strengths to take advantage of these opportunities?" Strengths and threats (ST) ask: "How can you take advantage of your strengths to avoid real and potential threats?" Weaknesses and opportunities (WO) ask: "How can you use your opportunities to overcome the weaknesses you are experiencing?" Weaknesses and threats (WT) ask: "How can you minimize your weaknesses and avoid threats?" An example of a SWOT analysis is shown in Figure 4.3.

Risks can be assessed using a risk assessment matrix (Figure 4.4). The risk impact assessment is a process of assessing the probabilities and consequences of risks if they occur. The results of a risk impact assessment will prioritize ranking from most to least critical with importance. Risks can be ranked in terms of their criticality to provide guidance on where resources should be deployed or mitigation practices should be put in place in order to reduce the risk.

After a process is mapped, a C&E diagram can be completed. This process is important because it provides for root cause analysis. The basis behind root cause analysis is to ask "why?" five times in order to get to the actual

Probability		Low	Medium	High
	High	Low impact high probability 3	Medium impact high probability 2	High impact high probability 1
	Medium	Low impact medium probability 4	Medium impact medium probability 2	High impact medium probability 2
	Low	Low impact low probability 4	Medium impact low probability 4	High impact low probability 3
		Low	Medium	High
			Impact	

FIGURE 4.4
Risk grid.

root cause. Often, problems are "band-aided" in order to fix the top-level problem, but the source of the problem is not addressed.

The C&E diagram is also referred to as a fishbone diagram, because it visually looks like a fish, where the bones are the causes and the fish head is the effect, as shown in Figure 4.5.

The fishbone is generally broken into the most important categories in a system, which include measurements, material, personnel (manpower), environment (Mother Nature), methods, and machines. These are also commonly referred to as the 6Ms. In addition, the headers can be named by performing an affinity diagram with the brainstormed ideas and using the theme for each grouping as the header for each fishbone. This process requires a team to do a great deal of brainstorming where they focus on the causes of the problems based on the categories. The "fish head" is the problem statement.

PERT analysis is an evaluation method that can be applied to time or cost. PERT provides a weighted assessment of time or cost. There are three key parameters for each scenario: optimistic (*O*) or best-case scenario, pessimistic (*P*) or worst-case scenario, and most likely (*ML*) scenario. From these parameters, the expected time, T_E, is calculated as (Equation 4.1)

$$T_E = \frac{O + 4ML + P}{6} \tag{4.1}$$

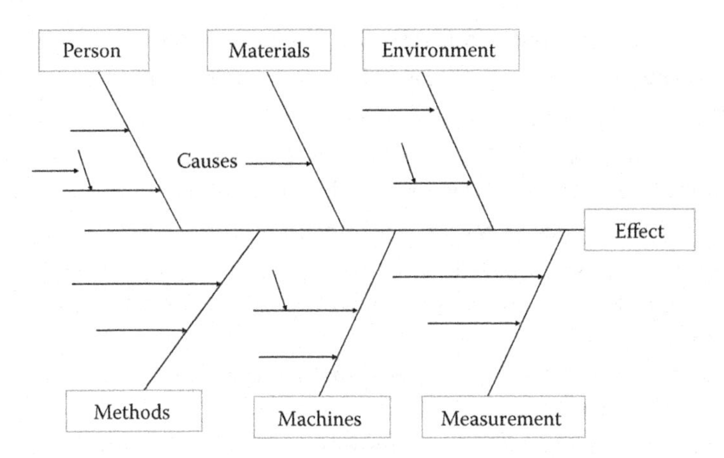

FIGURE 4.5
C&E diagram.

The expected time calculation provides for a heavier weighting of the *ML* scenario but also considers the best and worst cases. If the best- or worst-case scenario is an extreme situation, the PERT will account for it in a weighted manner.

Example

An organization needs to determine the expected time needed to receive a new computer. The supplier generally overestimates and has given a window of three weeks (21 days) for the *P* scenario. However, you have several coworkers who have received their computers in one week (seven days). This would be the *O* scenario. The *ML* scenario is the median delivery time of the supplier, which is 10 days. Calculate the expected time.

$$T_E = \frac{O + 4ML + P}{6}$$

$$T_E = \frac{7 + 4(10) + 21}{6}$$

$$T_E = 11.33 \text{ days}$$

Questions

1. What is status quo?
2. Why is change management important in launching a DFSS initiative?
3. Why do people fear change? Describe a situation that you have experienced in which you had to lead change. What resistance did you encounter?

4. Why is it important to address "what is in it for me?"
5. Describe three ways you can reduce resistance to change.
6. What are the required elements for change?
7. Describe three strategies for overcoming resistance. Provide an example scenario on how you would use those three strategies.
8. What are the success factors for implementing DFSS?
9. Describe five strategies for effectively communicating DFSS. Provide an example scenario on how you would use those five communication strategies.
10. Describe the five qualities that show it is time for a change. Give an example of each.
11. An organization needs to determine the expected time needed to design a new component. Based on past experience for a similar component, the longest time was 185 days, the average time was 127 days, and the shortest time was 73 days. Calculate the expected time.

References

Beckhard, R. and Pritchard, W. (1992). *Changing the Essence*. San Fransico, CA: Jossey-Bass.
Heller, R. (1999). *Managing Change*. London, United Kingdom: Dorling Kindersley.
Ricardo, R. (1995). Overcoming resistance to change, *National Productivity Review*, 14(2): 5.

Technical Design Review: Define and Measure Phases

The purpose of the Define phase in DFSS is to define the problem and the opportunity a new product, process, or service represents. In this phase, the team must identify customers and their product requirements.

In the Measure phase of DFSS, the team determines the baseline for the product, process, or service and gathers the data associated with the problem as well as the voice of the customer (VOC) data associated with the opportunity to design a new product, process, or service. The critical-to-quality (CTQ) flowdown to subsystems and components with specifications is performed. In addition, the team validates the measurement system by performing a gage repeatability and reproducibility (R&R) study and establishes the measurement systems analysis.

Technical Design Review

Technical design reviews apply a phase/gate approach to technology development. It is a structured and disciplined use of phases and gates to manage product commercialization. The purpose of a technical design review is to prevent design flaws and mistakes in the next phase of development and manage obstacles during a development phase.

A phase is a period of time during which specific tools and best practices are used in a disciplined manner to deliver tangible results that fulfill a predetermined set of technology development requirements. A gate is a stopping point within the flow of the phases in a DFSS project.

Two things occur at a gate:

1. A thorough assessment of deliverables from the tools and best practices that were conducted during the previous phase
2. A thorough review of the project management plans for the next phase

A three-tiered system of colors indicates the readiness of a DFSS project to pass through a gate:

1. **Green**: The subsystem has passed the gate criteria with no major problems.
2. **Yellow**: The subsystem has passed some of the gate criteria but has numerous moderate to minor problems that must be corrected for complete passage through the gate.
3. **Red**: The subsystem has passed some of the gate criteria but has one or more major problems that preclude the design from passing through the gate.

Checklists can also be used to list tools and best practices required to fulfill a gate deliverable within a phase. The checklist for each phase should be clear so that it is easy to directly relate one or more tools or best practices to each gate deliverable. Product planning can use the items in the checklist in the development of PERT or Gantt charts.

Scorecards are brief summary statements of deliverables from specific applications of tools and best practices. Deliverables should be stated based on the information to be provided, including who is specifically responsible for the deliverable and the due date. This helps the technical design review committee to manage the risk associated with keeping the project on schedule.

Gate 1 Readiness: Define and Measure Phases

- Proprietary, documented customer needs and technical building blocks

- Candidate concepts
- Candidate technical knowledge
- Well-planned commercial context
- Candidate technologies assessed

Assessment of Risks

The following risks should be quantified and evaluated against target values for this phase:

- Quality risks
- Delivery risks
- Cost risks
- Capital risks
- Performance risks
- Volume (sales) risks

5

Project Charter

Observing many companies in action, I am unable to point to a single instance in which stunning results were gotten without the active and personal leadership of the upper managers.

Joseph M. Juran

Introduction

A project charter is a definition of the project that includes the following:

- Problem statement
- Overview of scope, participants, goals, and requirements
- Provides authorization of a new project
- Identifies roles and responsibilities

Once the project charter is approved, it should not be changed.

A project charter begins with the project name, the department of focus, the focus area, and the product or process. A project charter serves as the focus point throughout the project to ensure that the project is on track and that the proper people are participating and are being held accountable.

The importance of a project charter with regard to sustainability is the living document to educate and give governance to a new project. Sustainability requires a great deal of education while providing goals and objectives. A project charter will serve as this living document for organizations with specified approaches.

Project Charter Steps

The key steps for developing a project charter include

1. Develop the problem statement
2. Identify the key customers and stakeholders

3. Develop the critical to satisfaction characteristics
4. Identify the project goals
5. Determine the project scope and products/processes/services to be improved
6. Identify the potential financial benefits
7. Determine potential project risks
8. Identify the project resources
9. Determine the project milestones

An example of a project template is provided in Figure 5.1. An example project charter is provided in Figure 5.2.

Risk Assessment

As with any project, there are risks. However, it is important to identify and take appropriate action early on to mitigate these risks. It is important that the

Project name: Provide a descriptive name that will be meaningful to non-team members.

Project overview: Provides a general description or background of the project.

Problem statement: Explain what the problem is in quantitative terms. The statement should also include what, when, impact, and consequences. This should be based on facts, for example, **4.7%** of steering knuckles fail prior to the end of their warranty of one year, which results in warranty costs of $278K in annual warranty costs in North America.

Customer/stakeholders: This should include both internal and external customers and stakeholders.

What is important to these customers: Critical to satisfaction (CTS), critical to quality (CTQ), critical to delivery (CTD) and critical to cost (CTC); examples include customer satisfaction, growth, delivery, every day low prices, waste reduction, environmental impact, etc.

Goal of the project: Explain what the improvement goal is in terms of percent improvement, improvement in sigma level, reduction in cost, increase in market share, etc.

Scope statement: Explain what is in and/or out of project scope; for example only Europe or Central US region.

Projected financial benefit(s): What do you think the estimated benefits to the business could be, could be cost avoidance, improved revenue, profit and loss (P&L) improvement, etc.

FIGURE 5.1
DFSS project charter template.

Project charter

Project name: Sure Mix Gasoline Can.

Team members: Joseph Baumann, Aiswarya Choppali, Sean Flachs, Jeffery Swanson.

Project overview: 2-cycle engines are prevalent in a majority of households across the nation. These engines are used in lawn mowers, leaf blowers, weed wackers, chain saws, boat engines, snowmobiles, and a multitude of other tools and devices. Specific to these engines, a precise mixing ratio of gasoline to 2-cycle engine oil, and in some cases marine 2-cycle oil, is required as specified by the product manual. Without an accurate ratio, consequences can range from reduced life span, reduced efficiency, engine failure, and ultimately increased risk to the user.

Problem statement: Customers are currently forced to utilize mixing cups to add specified amounts of 2-cycle oil to their gas cans. These measuring devices must not only be collocated and carried with the gas can but also must be cleaned, which adds to an environmentally unfriendly hazmat disposal issue. Further, use of these measuring cups runs the possibility of adding debris to the mixture ultimately damaging the internal parts of the engine. Critical to success is introducing a method of incorporating a measuring or mixture method that does not utilize a separate object/tool and reduces the possibility for debris and other outside materials to enter the gas can while still providing durability, usability, and longevity to the customer.

Customer/stakeholders: The target customer base is males, ages 22–65, which own lawn and/or marine 2-cycle equipment. Further, household income should be within the $35K to $250K to target a populous that would be capable of expending slightly more capital for convenience and accurate capabilities of a simple, but highly utilized product.

Goal of the project: The project goal for the Sure Mix Gasoline Can is to eliminate the need for additional measuring devices allowing the customer to purchase 2 cycle oil at any quantity, reduce the hazmat footprint by more than 50%, and increase customer confidence in mixture accuracy by 25%.

Scope statement: The scope of this project will remain within the United States; more specifically a trial period would be expected to be introduced following approval of a lead design product within the Midwest Region of the United States. This area has been chosen due to the close proximity for supply chains as well as higher than average use of 2-cycle engines.

Projected financial benefit(s): This product has the potential to allow the customer base to reduce overall cost and increase the life span of their 2-cycle engine products. This cost savings can be expected to be the driving force towards the purchase and desire to own the new product. Consumers have shown over time that they are willing to spend slightly more on the front end for expected benefits on the back end. Reduced cost for measuring cups, cleanup operation & disposal, and providing the ability to purchase 2-cycle oil in larger quantities with cost savings supplemented by convenience and confidence in mixing their gasoline will motivate them to choose this product over ordinary gasoline cans. Expected outcomes will be increased market share and revenue within the small gas can arena.

FIGURE 5.2
Example DFSS project charter.

team understands the potential risks for the project to be successful. As part of the project charter, the team can identify the risks at the start of the project. If this is not included as part of the project charter, the team can document the potential risks at the first team meeting and review the risks at subsequent team meetings. Figure 5.3 provides a template for risk assessment.

Potential risks	Probability of risk (H/M/L)	Impact of risk (H/M/L)	Risk mitigation strategy

FIGURE 5.3
Risk assessment.

Potential risks are the events that could occur and, if they occur, would affect the success of the project. The probability of risk is rated as high, medium, or low (H/M/L) and represents the risk that would occur. The impact of the risk uses the same H/M/L scale and indicates the risk to the customer. Finally, based on the risk, the team should determine how to reduce or eliminate the impact of the risk through the risk mitigation strategy. This enables the team to be proactive early in the project by determining contingency plans and/or what-if scenarios. By creating this risk assessment plan/strategy, the team can help ensure the success of the project.

Developing the Business Case

Justification should be sought when making an investment of any kind in a project. This entails a comparison of industry objectives with the overall national goals in the areas discussed here.

Market target: This should identify the customers of the proposed technology. It should also address items such as the market cost of the proposed product, an assessment of competition, and market share.

Growth potential: This should address short-range expectations, long-range expectations, future competitiveness, future capability, and the prevailing size and strength of the competition.

Contributions to sustainability goals: Any prospective technology must be evaluated in terms of the direct and indirect benefits to be generated by the technology. These may include product price versus value, increase in international trade, improved standard of living, cleaner environment, safer workplace, and improved productivity.

Profitability: An analysis of how the technology will contribute to profitability should consider the past performance of the technology, incremental benefits of the new technology versus conventional technology, and value added by the new technology.

Capital investment: A comprehensive economic analysis should play a significant role in the technology assessment process. This may cover an evaluation of fixed and sunk costs, cost of obsolescence, maintenance requirements, recurring costs, installation cost, space requirement cost, capital substitution potentials, return on investment, tax implications, cost of capital, and other concurrent projects.

Skill and resource requirements: The utilization of resources (manpower and equipment) in the pre- and post-technology phases of industrialization should be assessed. This may be based on material input/output flows, high value of equipment versus productivity improvement, required inputs for the technology, expected output of the technology, and utilization of technical and nontechnical personnel.

Risk exposure: Uncertainty is a reality in technology adoption efforts. Uncertainty will need to be assessed for the initial investment, return on investment, payback period, public reactions, environmental impact, and volatility of the technology.

National sustainability improvement: An analysis of how the technology may contribute to national sustainability goals may be verified by studying industry throughput, efficiency of production processes, utilization of raw materials, equipment maintenance, absenteeism, learning rate, and design-to-production cycle.

Conclusion

A project charter outlines the problem statement and provides an overview of the scope, participants, goals, and requirements for the Design for Six Sigma (DFSS) project. It is a critical part of any Six Sigma or DFSS project because it identifies the roles and responsibilities within the team and provides authorization for new projects since the team members and upper management sign the project charter. This ensures that the project is on track and the proper people are participating and being held accountable. Chapter 6 discusses the balanced scorecard, which enables the team to select the appropriate, holistic metrics in terms of the organization's overall vision and strategy when developing a new product, process, or service.

Questions

1. Why is a risk assessment important in a project charter?
2. What are the steps in developing a project charter?
3. Which portion of the project charter provides a general description of the project?
4. Give an example of a project description and the associated key customers and stakeholders.
5. What is the purpose of the scope in the project charter?
6. Why is the business case important?
7. What information should be included in the business case?

6

Balanced Scorecard

0 plus 100 equals 100. But so does 50 plus 50, only with more balance.

Jarod Kintz

The business case should be identified early in the project and clearly outlined in the project charter (covered in Chapter 5). The business results should be related to the business, project, or process and are often referred to as the performance measures. As the project progresses, the business results should be tracked and refined. One of the main methodologies for developing the business results is the balanced scorecard.

Balanced Scorecard

Drs. Robert Kaplan and David Norton of the Harvard Business School developed the balanced scorecard. The methodology was developed to align the business activities within an organization with the overall vision and strategy and provide focus on the strategic agenda. The alignment improves internal and external communication and aids in monitoring performance against the strategic goals by selecting a small number of critical metrics for the organization. These metrics include a balance of financial and nonfinancial metrics.

The balanced scorecard consists of four key perspectives (financial, customer, internal business process, and learning and growth), which provide a balanced view of the organizational performance. The balanced scorecard is shown in Figure 6.1.

The financial perspective focuses on how well the goals, strategies, and objectives contribute to the bottom line and stakeholder value. This perspective involves the traditional financial data, which can provide timely and accurate information on the health of an organization. However, this can occasionally lead to an unbalanced situation by focusing solely on financial information. Therefore, additional financial-related data such as risk assessment and cost–benefit analysis also falls into this perspective. Examples of common financial metrics include return on investment, return on capital,

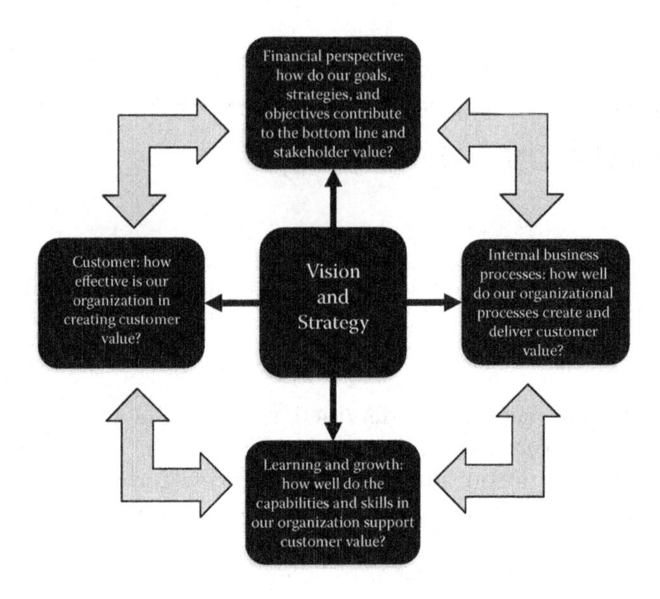

FIGURE 6.1
Balanced scorecard.

return on equity, inventory turns, gross margin, economic value added, cost of poor quality, and cost savings.

Customer perspective is used to determine how effective the organization is in creating value for its customers. This perspective focuses on the customer and customer satisfaction; therefore, the metrics should represent the various customer segments and the processes that produce or deliver their product or service. Common examples of customer metrics include customer satisfaction rate, customer retention, referral rates, quality, on-time delivery, and number of new products launched this year.

The third perspective, internal business process, is used to measure how well the organizational processes are able to create and deliver the customer's value proposition. Metrics for this perspective should be selected to indicate how well the business is running and whether the products and services meet customer requirements and expectations. Common internal business process metrics include defect rates, cycle time, throughput rates, lead time, uptime, and non-value-added ratio.

The learning and growth perspective addresses how well the capabilities and skills within the organization are focused on supporting internal processes and creating customer value. This includes employee training and corporate culture. Particularly with rapid technological advances, continuous learning is essential within organizations; it can be tracked through training hours and other metrics. It is important to note, though, that learning involves more than training; it also includes mentoring and technological tools. Common learning and growth metrics include employee turnover, employee morale/satisfaction, percentage of internal promotions, percentage of succession plans completed, absenteeism, and number of employees trained as a Six Sigma Black Belt or Green Belt.

There are four key steps for creating a balanced scorecard. In creating the balanced scorecard for an organization, all steps should flow strategically together and have a clear linkage.

Step 1: The first step is to identify the organization's key strategies and mission and translate the vision into operational goals for the organization. The executive team typically sets the strategies and mission based on the organization's key competencies and competition.

Step 2: The specific goals and objectives that need to be accomplished to support the strategic plan for the organization are then determined and communicated throughout the organization. This helps link the vision to individual performance.

Step 3: The metrics for each level of the company are then determined, and targets are set for business planning. These metrics should link directly to the organizational strategic goals and objectives.

Step 4: The final step is feedback and learning. Continuous improvement and communication are used to adjust the strategy as appropriate.

Key Performance Indicators

To have a balanced view of an organization, a balanced set of metrics is necessary, similar to a dashboard. This enables an overall picture of the organizational performance and provides holistic and balanced goals for all employees. A balanced set of metrics includes leading and lagging indicators. There is a cause and effect relationship between the objectives and strategies in the balanced scorecard. Likewise, there is a cause and effect relationship between leading and lagging indicators. For example, satisfied and motivated employees are a leading indicator of customer satisfaction. Lagging indicators are the traditional business performance metrics such as unit manufacturing costs, defects per million opportunities (DPMO), complaints received, and mean time to failure, and are often used to measure output. Lagging indicators typically provide data after something has occurred, and can be the result of changes in leading indicators. Leading indicators are used to predict a future event, and can be used to drive and measure continuous improvement activities. Leading indicators are proactive and are often captured at the process or activity level; however, they may not always be accurate, since they are predictions. Examples of leading indicators include error rates and on-time delivery. These measure functional performance and include metrics such as physical functions.

Key performance indicators (KPIs) involve both financial and nonfinancial metrics. The criteria for a KPI is that it should be quantitative and measurable, goal based, process based, strategy based, and time bounded. These are commonly referred to as SMART metrics, which stands for

- Specific: Focused and process based
- Measurable: Quantitative and easily determined
- Achievable: Obtainable
- Relevant: Linked to the organization's strategies, goals, and objectives
- Time bounded: Specific time period

There are numerous measures that focus on the project, process, and financial performance.

Cost of Quality

Cost of quality is a methodology that is used to express quality in monetary terms. There are four categories of quality costs: prevention, appraisal, internal failure, and external failure, as shown in Figure 6.2.

Prevention costs are the costs incurred to prevent defects. Examples of these costs include new product review, process control, improvement projects, quality planning, training for quality, quality system development, and quality information systems.

Appraisal costs are those costs incurred during inspection. Examples of these costs include inspecting incoming material, material in inventory, material in process, and finished product. Since inspection costs are covered in this category, the costs of maintaining the integrity of instruments and gauges are also included as appraisal costs.

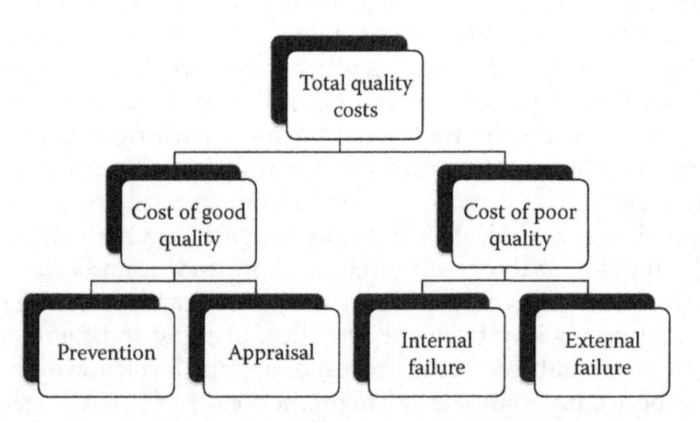

FIGURE 6.2
Cost of quality categories.

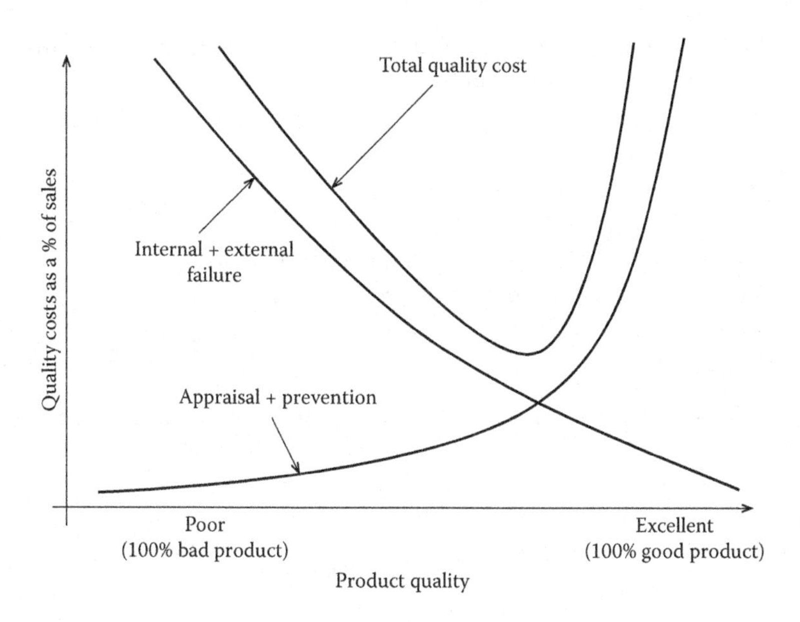

FIGURE 6.3
Cost of quality expenditures.

Internal failure costs are the costs due to failures internally. Examples of these costs include scrap, rework, retest, and penalties for not meeting customer schedules.

External failure costs are those costs due to defective products reaching the customer. Examples of these costs include warranty charges, defect campaigns, complaint adjustments, and returned materials.

Financial resources should be balanced between these categories, and to achieve high levels of quality, financial resources should be focused on appraisal and prevention, as shown in Figure 6.3.

Financial Performance

In terms of project performance, there are two key measures: cost performance index (CPI) and schedule performance index (SPI). CPI is used to measure the project performance in monetary terms. CPI is the ratio of the value earned to the actual cost of the work performed. SPI is used to measure the efficiency of a project's schedule. SPI is a ratio of the earned value to the planned value.

Several additional financial measures include revenue growth, market share, margin, and net present value. Revenue growth is a measure of the projected increase in income for the organization as a result of completing a project. Market share represents the percentage of the dollar value sold relative to the dollar value sold by all other organizations within that given market. Margin is calculated as the difference between the income and the cost. Net present value is the amount that will be received in the future. It is calculated as shown in Equation 6.1.

$$P = A(1+i)^{-n} \tag{6.1}$$

where:

P = net present value
A = amount to be received n years from now
i = interest rate expressed as a decimal

Example

Assume that in five years, $12,500 will be available. Assuming an annual interest rate of 6%, what is the net present value of that money?

$$P = A(1+i)^{-n}$$

$$P = \$12,500(1+0.06)^{-5}$$

$$P = \$9,340.73$$

Therefore, $9,340.73 invested at 6% compounded annually will be worth $12,500 after five years.

Process Performance

Process performance uses measures that determine how a process is operating against established goals or statistical measures. The most common process performance metrics include defects per unit (DPU), DPMO, rolled throughput yield (RTY), sigma level, and process capability indices.

DPU is a ratio of the total number of defects to the total number of products produced during a given time. DPU is shown in Equation 6.2.

$$\text{DPU} = \frac{\text{Total number of defects}}{\text{Total number of products produced (good and bad)}} \tag{6.2}$$

Example

A manufacturer of control valves produced 17 defective parts and 333 good parts in one production run. Calculate the DPU.

$$\text{DPU} = \frac{\text{Total number of defects}}{\text{Total number of products produced (good and bad)}}$$

$$\text{DPU} = \frac{17}{333+17}$$

$$\text{DPU} = 0.049$$

DPMO is calculated by dividing the total number of defects by the total number of opportunities. To calculate DPMO, the number of ways a defect can occur for each item must be known. The calculation for DPMO is given in Equation 6.3.

$$\text{DPMO} = \frac{\text{No. of defects}}{\text{No. of parts produced} \times \text{No. of opportunities}} \times 1{,}000{,}000 \tag{6.3}$$

Example

A manufacturing company produces axle shafts for small trucks. Each axle shaft has a milled roller bearing surface. This surface has four characteristics or opportunities: length, diameter, surface finish, and roundness. Through inspection, a total of 312 defects were found out of a total production of 18,000 axles. Calculate the DPMO.

$$\text{DPMO} = \frac{\text{No. of defects}}{\text{No. of parts produced} \times \text{No. of opportunities}} \times 1{,}000{,}000$$

$$\text{DPMO} = \frac{312}{(18{,}000)(4)} \times 1{,}000{,}000$$

$$\text{DPMO} = 4{,}333$$

RTY is a measure of the yield from a series of processes. It is calculated by multiplying the individual process yields, as shown in Equation 6.4.

$$\text{RTY} = Y_1 \times Y_2 \times \cdots \times Y_n \tag{6.4}$$

Example

A product goes through four processes. The yields for the four process steps are 0.994, 0.987, 0.951, and 0.990. Calculate the RTY.

$$\text{RTY} = Y_1 \times Y_2 \times \cdots \times Y_n$$

$$\text{RTY} = 0.994 \times 0.987 \times 0.951 \times 0.990 = 0.924$$

A Six Sigma process can fit six standard deviations between the mean of the process and the closest specification limits, as shown in Figure 6.4.

Process capability indices are dimensionless numbers. These indices are used to compare the variability of a characteristic with the specification limits. There are three main indices: C_p, C_{pk}, and C_r.

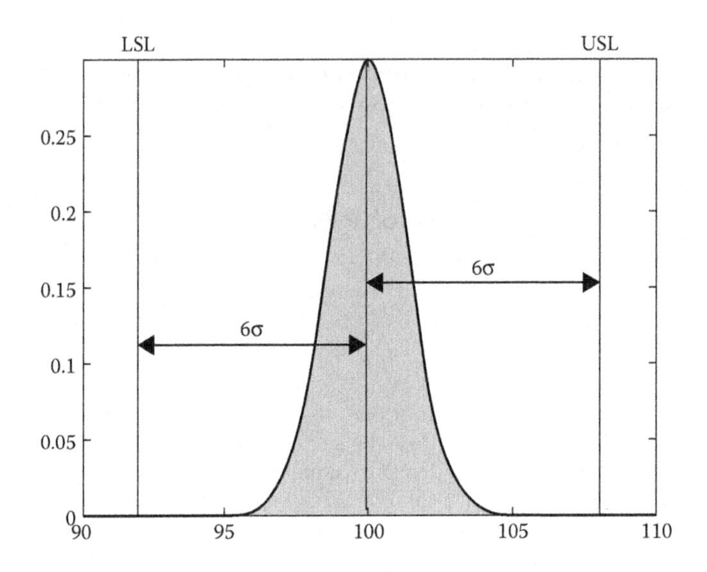

FIGURE 6.4
Sigma level.

The C_p index is an estimate of the process capability if the process mean is centered between the specification limits. It assumes the output of the process is normally distributed. The C_p calculation is shown in Equation 6.5.

$$C_p = \frac{\text{USL} - \text{LSL}}{6\sigma} \tag{6.5}$$

where:
USL upper specification limit
LSL lower specification limit
σ natural process variation

Equation 6.6 is another perspective of the calculation for C_p.

$$C_p = \frac{\text{Voice of the customer}}{\text{Voice of the process}} = \frac{\text{VOC}}{\text{VOP}} = \frac{\text{Tolerance}}{\text{Natural process variation}}$$

$$= \frac{\text{USL} - \text{LSL}}{6\sigma} \tag{6.6}$$

A C_p of 1.0 is technically capable; however, since C_p does not take centering into consideration, this does not mean that the process is acceptable. A C_p of 2.0 represents 6σ performance.

C_{pk} was developed to address the issue of process centering. This index measures the proximity of the process mean, μ, to the nearest specification limit, as shown in Equation 6.7. It also assumes that the output of the process is normally distributed.

$$C_{pk} = \min\left[\frac{USL-\mu}{3\sigma}, \frac{\mu-LSL}{3\sigma}\right] \tag{6.7}$$

Common industry goals for C_{pk} are 1.33 or 1.67, since they allow some process drift. The C_{pk} value can be converted to the sigma level, as shown in Equation 6.8. Therefore, a C_{pk} of 2.0 is at the Six Sigma level.

$$\text{Sigma level} = 3C_{pk} \tag{6.8}$$

C_r is the capability ratio, which is the inverse of C_p. For the capability ratio, C_r, the lower the value, the better.

$$C_r = \frac{1}{C_p} \tag{6.9}$$

Example

A Design for Six Sigma team in the light truck industry is redesigning a gear housing. Prior to redesigning the product, the team wants to understand the current process capability of a critical bore. The critical bore has specifications of 1.120–1.125 inches. A control chart shows the process is stable. The standard deviation of the process is 0.0011, and the sample mean is 1.1221. Determine C_p, C_{pk}, and C_r.

$$C_p = \frac{USL-LSL}{6\sigma}$$

$$C_p = \frac{1.125-1.120}{6(0.0011)} = \frac{0.005}{0.0066}$$

$$C_p = 0.758$$

Since the C_p value is less than 1.0, the process is not capable. This indicates that the process width (variation) is greater than the specification width.

$$C_{pk} = \min\left[\frac{USL-\mu}{3\sigma}, \frac{\mu-LSL}{3\sigma}\right]$$

$$C_{pk} = \min\left[\frac{1.125-1.1221}{3(0.0011)}, \frac{1.1221-1.120}{3(0.0011)}\right]$$

$$C_{pk} = \min[0.879, 0.636]$$

$$C_{pk} = 0.636$$

The values for the C_{pk} are not equal; therefore, the process is not centered.

$$\text{Sigma level} = 3C_{pk}$$

$$\text{Sigma level} = 3(0.636)$$

$$\text{Sigma level} = 1.908$$

$$C_r = \frac{1}{C_p}$$

$$C_r = \frac{1}{0.758}$$

$$C_r = 1.319$$

Based on the capability indices, the process contains high variation and is not centered.

Conclusion

The balanced scorecard can drive the selection of appropriate KPIs, which can result in a positive impact on the overall organizational performance by linking the metrics to the overall vision and strategy of the organization. It provides a balanced and holistic approach between business performance, continuous learning, and value to customers while providing long-term financial growth for an organization. By communicating the vision and strategy and driving the communication and metrics throughout the organization, this also provides a clear direction for continuous improvement.

Questions

1. Describe each of the balanced scorecard perspectives.
2. Give an example metric for each of the balanced scorecard perspectives.
3. In the balanced scorecard approach, what perspective addresses the capabilities and skills of an organization, and how is it focused to support internal processes and create customer value?
4. A manufacturing company produces axle shafts for small trucks. Each axle shaft has a milled roller bearing surface. This surface has four characteristics or opportunities: length, diameter, surface finish, and roundness. Through inspection, a total of 312 defects were found out of a total production of 18,000 axles. Calculate the DPMO.

5. Given a C_{pk} value of 1.5, what is the sigma level?
6. Determine the C_p value for a diameter with an upper specification limit of 2.50 and a lower specification limit of 2.25, given the following sample data:

2.12	2.43	2.28	2.37	2.31

7. Calculate the process capability, C_p, given the following information: USL = 1.755, LSL = 1.745, $\sigma = 0.0017$. Comment on the process. What can you say about the process in terms of centering and variation?
8. Determine the C_p value for an inner bore diameter with an upper specification limit of 1.50 and a lower specification limit of 1.25, given a standard deviation of 0.23. What can you interpret about the process?
9. Using the same data as in Problem 6 and a mean of 1.35, calculate the C_{pk}. What can you interpret about the process?
10. A sample of 55 measurements is taken from a process with a standard deviation of 0.00213 and a mean of 2.23. The product has specification limits of 2.20 and 2.25. Calculate the value of C_{pk}. Assume the process is normal. What can you say about the process in terms of centering and variation?
11. In March, several engineers visited a computer manufacturer in Sacramento to help them improve their process capability for manufacturing a precision ring used on high-speed disk drives. The specifications for the outer diameter of the ring are 1.525 ± 0.001 inches, which is a critical dimension. A capability analysis of the process before improvement techniques yielded the following statistics from 86 runs: mean of 1.5257 and standard deviation of 0.000293. Calculate the process capability indices, C_p and C_{pk}.
12. A sample of 100 measurements was taken from a process with a sample standard deviation of 0.12. The product has specification limits at 24.50 and 25.50. Calculate the values of C_p and C_{pk} if the process average is 25.04. What can you say about the process in terms of centering and variation?

7

Benchmarking

If your actions inspire others to dream more, learn more, do more and become more, you are a leader.

John Q. Adams

Think about the first immaculate piece of technology you worked with. Shortly after, were there very similar pieces of that same technology, but maybe even a little bit better? How did that happen? Utilizing best-in-class (BIC) practices, better and better items are made. A person takes the BIC material, determines exactly how to produce it, and then essentially makes it better. If the first BIC product is not replicated, there would be a monopoly of companies being the only company in the business making that particular product. Each company now takes the BIC products and continuously improves them in order to stay competitive in the marketplace. If each company does not continuously improve or is not innovative, the competition will take the business and the customers will follow quickly.

Best in Class

A best practice is a technique or methodology that, through experience and research, has proved to reliably lead to a desired result. A commitment to using the best practices in any field is a commitment to using all the knowledge and technology at one's disposal to ensure success.

In today's economic market, if you do what you have always done, you will get what you have always got. Customers may currently be happy with the product or service, but once they find an innovative new product or service, they are automatically drawn to it no matter their loyalty. Adopting philosophies and methodologies that make the business BIC is what makes a competitive business.

Utilizing new technology in the business will help the business grow by being more efficient and innovative. The systems being used help the business grow and the benefits of the new technologies attract new customers.

Communicating with key customers and suppliers is important so that they are in tune with changes and understand the need for the business to keep growing and the desire to be BIC. This helps relationships to stay healthy at both ends and avoids misrepresentation.

The business should use a total-cost approach versus a price-of-product approach. This includes all aspects of the business. Therefore, labor and materials are not the only overheads, but the entire supply chain is looked at. What sourcing provides the highest quality at the lowest cost, what methods of transportation can be used to reduce costs, and what bulk quantities must not be overproduced to keep the customers happy? Operating, training, maintenance, warehousing, environmental, and quality costs should be under this umbrella as well. Working with suppliers, the organization must determine how to reduce the costs for both businesses in order to stay BIC in the overall business that is being conducted. Controlling processes is an essential part to being BIC because the process is standardized and validated for profit. Risk should be eliminated or mitigated so that unexpected events do not occur that increase costs more than normal operating practices. Identifying and eliminating risk should include

- Identifying areas of risk
- Determining probability, occurrence, and detection of risk
- Determining costs of risk
- Prioritizing risk for monitoring and prevention

Best practices spread through companies and fields of interest after a success occurs. Skills and knowledge are key practices to grow and share in order to enable BIC techniques. BIC is also a technique that is considered to yield consistently superior results; therefore, benchmarking techniques are utilized to encourage leading practice results.

Benchmarking endeavors involve strategic and analytical planning. It is a methodology for comparing the current state with the BIC products or processes. The competitive strategy is to acquire knowledge of the BIC process or product by comparing the processes and understanding the methodology. Benchmarking enables teams to set targets and determine ways to achieve these targets. Once the process is understood and compared for differences, the process should be made more efficient, with a similar methodology, but without copying the exact product. If any items do not make sense in the processes, they should be eliminated or improved so the new product or service is the new BIC technique.

There are four main types of benchmarking practices:

1. Internal benchmarking
2. External benchmarking
3. Functional benchmarking
4. BIC practices

Internal benchmarking seeks current practices within the business and improves its own functions and processes. Any processes that are better than others are looked at for the BIC approaches and those approaches are used for new processes. External benchmarking is a comparison of a business that makes similar products. This benchmarking is used not to copy the other product or service, but instead to use the innovative ideas that are attractive to customers to improve one's own product or service. External benchmarking is normally completed with a competitor. Functional benchmarking focuses on the same functions within the same industries or businesses to be competitive without severe analytical or strategic skills. It takes high-level BIC processes and gives customers the same advantage. Generic benchmarking takes BIC practices across any organization regardless of if it is the same or not and utilizes the attractive aspects to make their own internal processes better.

Strategic planning utilizes project management techniques to dissect technology in order to make a better business plan. It utilizes the goals, visions, and strategies of the business to ensure alignment with the best in the business.

Analyses must be conducted during strategic planning so that the entire process is mapped out without jumping to improvement measures before the definition and measurements have taken place.

Benchmarking and BIC practices come from communication from the customer. The customer is the number one priority, so feedback from the customer must be obtained in order to understand what it is the customer wants. If a new label is put on yearly to attract the customer, the customer may be attracted, but truly not care about the label and care more about the quality of the product. A customer analysis should be conducted to implement benchmarking of the right opportunities.

The speed at which the BIC practices take place is important because customers do not want to wait to receive their items. They want rush delivery on the best quality service they can find at a competitive price. If they do not get what they desire, they will go to another business without even thinking twice about the business that is making them wait.

Findings must be shared in order for the strategy to continue. Working outside the office helps with strategy because it is seeking other endeavors or products to innovate. Most of the time, ideas spark from brainstorming other products or services to find new products or services that are different in nature, but represent what the customer is looking for. This balance is depicted in Figure 7.1.

The Best Manufacturing Practices (BMP) Program encourages companies to operate at a higher level of efficiency and become more competitive through the implementation and sharing of best practices. The BMP Program was established to enable the U.S. defense industry and the U.S. industrial base to improve product price, quality, and delivery. These types of practices can be found in any country and are encouraged as a means of building competitive advantages for companies.

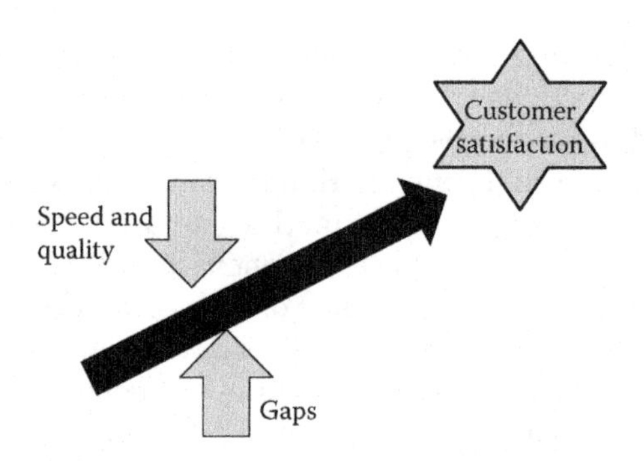

FIGURE 7.1
Speed and quality vs. gaps.

Finding the gap between your particular business sector and the competitor is a start to understanding the amount of work it will take to get to the BIC process. Understanding key business metrics and standard operating procedures is also one of the first steps in finding ways to get to BIC practices. Understanding where to begin focus-improvement projects is a means to improving so that the project stays in scope and is actually attainable. As organizations begin to strive to attain BIC, performance automatically begins to increase, as shown in Figure 7.2.

The operational performance of a business is based on the solid performance of that business and on its structure. If this strategy is properly analyzed, the business can be BIC in any operation it desires (Figure 7.3).

To ensure the business is operating at a model desirable to the company, scorecards for BIC practices can be conducted. These include an evaluation of the following key metrics:

- Quality practices
- Quality performance
- Lean manufacturing practices
- Cost metrics

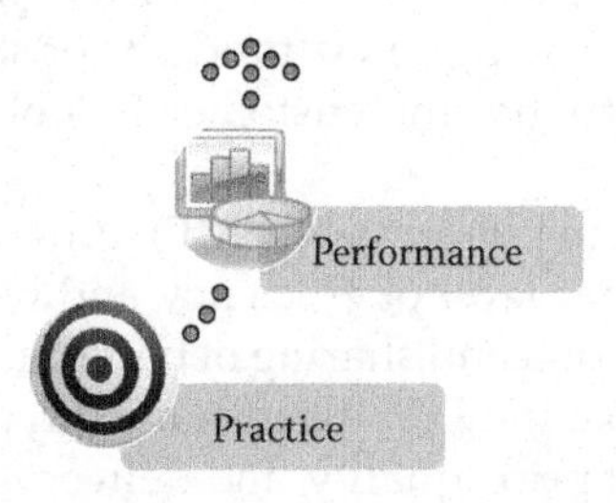

FIGURE 7.2
Practice vs. performance.

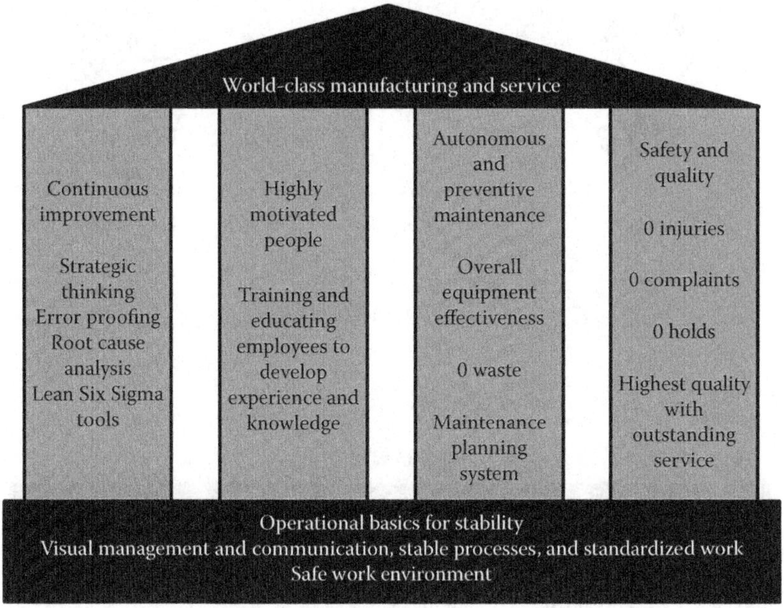

FIGURE 7.3
Operational stability model.

- Supply chain performance
- Customer satisfaction
- Leadership and communication skills

The amount of resources put into these metrics will ensure that the business stays competitive in the marketplace.

The company view should include the SIPOC model, which shows (Figure 7.4)

- Suppliers: Internal and external
- Inputs: Feed the process
- Processes: Steps or sequence of steps
- Outputs: Measureable product or service
- Customers: People who receive the output

An example of a SIPOC diagram is provided in Figure 7.5 for the process of making frosting.

The company view involves understanding the processes first and foremost. Processes are management's responsibility: management must understand all processes and the rationale behind the processes. The inputs and

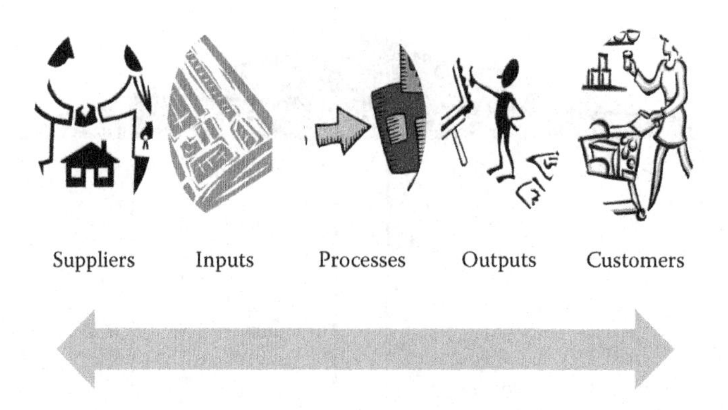

FIGURE 7.4
SIPOC model.

Suppliers	Inputs	Processes	Outputs	Customers
ABC	Soy bean oil	Put into slurry kettle		
Cargo	Shortening 112	Put into slurry kettle		
Lodi	Emulsifier	Put into slurry kettle		
Breen	Polysorbate	Put into slurry kettle		
Soso	Lecitin	Put into slurry kettle		
		Heat to 180°		
ABC	Corn syrup	Put into slurry kettle	Slurry mix	Icing blender
Cargo	Water	Put into slurry kettle		
Cargo	Salt	Put into slurry kettle		
Breen	Sorbic acid	Put into slurry kettle		
Western	Granulated sugar	Put into slurry kettle		
		Blend 5 minutes and hold at 160°		
Bird	Cocoa	Add cocoa		

FIGURE 7.5
SIPOC example.

outputs of the processes are looked upon next to ensure that there are no gaps and that the correct output is being produced. Finally, supplier and customer relations are an essential part of BIC practices, and must be present in order for processes to improve and succeed.

Conclusion

Benchmarking is a strategic and analytical methodology for comparing the current state with the BIC products or processes. Understanding key business metrics and standard operating procedures also enables teams to set targets and determine ways to achieve these targets and become BIC. Once the necessary steps to achieve BIC are determined, the team should use project management techniques to manage the steps. Project management is covered in Chapter 8.

Questions

1. A bank has decided that they need to improve their loan process. Therefore, the bank forms a team to compare their process with a similar operation to identify opportunities for improvement. The team decides to benchmark an insurance application at a life insurance company since it has a similar process. Which form of benchmarking does this reference?
2. From whose perspective should benchmarking be performed in product design?

Reference

Agustiady, T. and Badiru, B. (2012). *Statistical Techniques for Project Control*. New York: CRC Press.

8

Project Management

If you don't know where you are going, how can you expect to get there?

Basil S. Walsh

Project management is the process of managing, allocating, and timing resources to achieve a given goal in an efficient and expeditious manner. The objectives that constitute the specified goal may be in terms of time, cost, or technical results. A project can be simple or complex. In each case, proven project management processes must be followed. In all cases of project management implementation, control must be exercised to ensure that project objectives are achieved.

Project management, which relies on knowledge acquired in this way, is the pursuit of organizational goals within the constraints of time, cost, and quality expectations. This can be summarized by a few basic questions:

- What needs to be done?
- What can be done?
- What will be done?
- Who will do it?
- When will it be done?
- Where will it be done?
- How will it be done?

The factors of time, cost, and quality are synchronized to answer these questions and must be managed and controlled within the constraints of the iron triangle depicted in Figure 8.1. In this case, quality represents the composite collection of project requirements. In a situation in which precise optimization is not possible, trade-offs among the three factors of success are required. A rigid (iron) triangle of constraints encases the project (Figure 8.1). Everything must be accomplished within the boundaries of time, cost, and quality. If better quality is expected, a compromise along the axes of time and cost must be executed, thereby altering the shape of the triangle.

The trade-off relationships are not linear and must be visualized in a multidimensional context. This is better articulated by a three-dimensional (3-D) view of the system's constraints, as shown in Figure 8.2. Scope requirements determine the project boundary, and trade-offs must be achieved within that

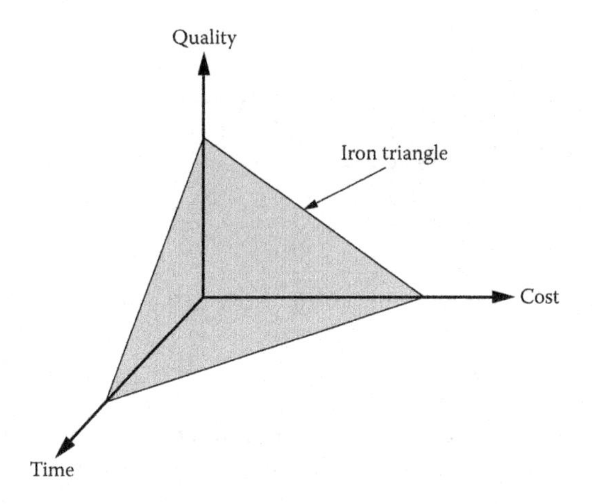

FIGURE 8.1
Project constraints of cost, time, and quality.

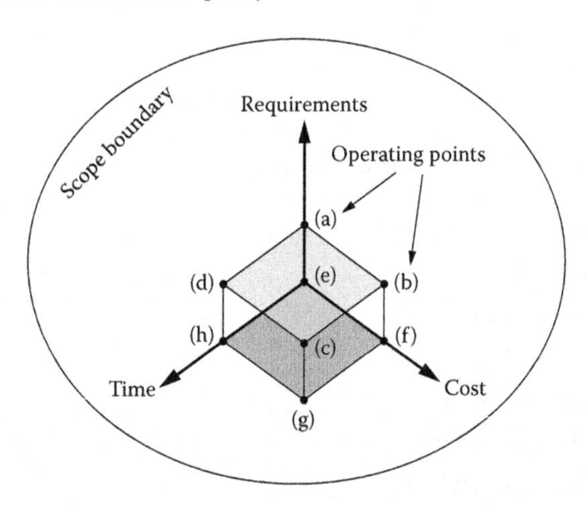

FIGURE 8.2
Systems constraints of cost, time, and quality within iron triangle.

boundary. If we label the eight corners of the box as (a), (b), (c),…, (h), we can iteratively assess the best operating point for a project. For example, we can address the following two operational questions:

1. From the point of view of the project sponsor, which corner is the most desired operating point in terms of combination of requirements, time, and cost?
2. From the point of view of the project executor, which corner is the most desired operating point in terms of combination of requirements, time, and cost?

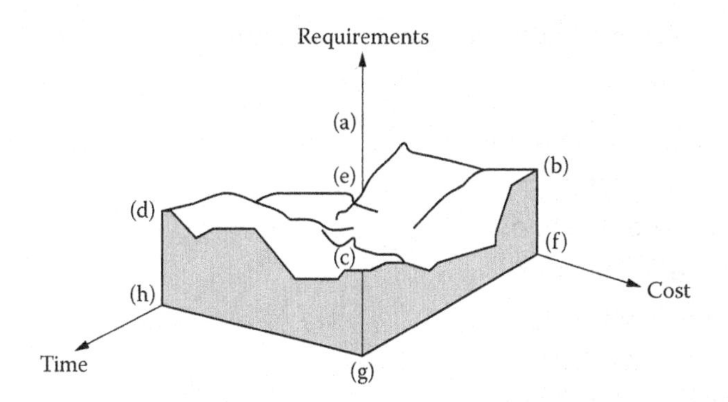

FIGURE 8.3
Compromise surface for cost, time, and requirements trade-off.

Note that all the corners represent extreme operating points. We notice that point (e) is the do-nothing state, where there are no requirements, no time allocations, and no cost incurrence. This cannot be the desired operating state of any organization that seeks to remain productive. Point (a) represents an extreme case of meeting all requirements with no investment of time or cost allocation. This is an unrealistic extreme in any practical environment. It represents a case of getting something for nothing. Yet, it is the most desired operating point for the project sponsor.

By comparison, point (c) provides the maximum possible for requirements, cost, and time. In other words, the highest levels of requirements can be met if the maximum possible time is allowed and the highest possible budget is allocated. This is an unrealistic expectation in any resource-conscious organization. You cannot get everything you ask for to execute a project. Yet, it is the most desired operating point for the project executor. Considering the two extreme points of (a) and (c), it is obvious that a project must be executed within some compromise region inside the scope boundary. Figure 8.3 shows a possible compromise surface with peaks and valleys representing give-and-take trade-off points within the constrained box. With proper control strategies, the project team can guide the project in the appropriate directions. The challenge is to devise an analytical modeling technique to guide decision-making over the compromise region. If we could collect sets of data over several repetitions of identical projects, we could model a decision surface to guide future executions of similar projects.

Why Projects Fail

Despite concerted efforts to maintain control of a project, many projects still fail. To maintain better control of a project, albeit not perfect control, we must understand the common reasons why projects fail. With this knowledge, we can better preempt project problems. Some common causes of project failure are as follows:

- Lack of communication
- Lack of cooperation
- Lack of coordination
- Diminished interest
- Diminished resources
- Change of objectives
- Change of stakeholders
- Change of priority
- Change of perspective
- Change of ownership
- Change of scope
- Budget cut
- Shift in milestone
- New technology
- New personnel
- Lack of training
- New unaligned capability
- Market shift
- Change of management philosophy
- Loss of manager (moves on)
- Depletion of project goodwill
- Lack of user involvement
- Three strikes followed by an out (too many mistakes)

Management by Project

Project management continues to grow as an effective means of managing functions in any organization. Project management should be an enterprise-wide systems-based endeavor. Enterprise-wide project management is the application of project management techniques and practices across the full scope of an enterprise. This concept is also referred to as management by project (MBP), an approach that employs project management techniques in various functions in an organization. MBP recommends pursuing endeavors as project-oriented activities. It is an effective way to conduct any business activity. It represents a disciplined approach that defines each work assignment as a project. Under MBP, every undertaking is viewed as a project that must be managed just like a traditional project. The characteristics required of each project include

- Identified scope and goal
- Desired completion time
- Availability of resources
- Defined performance measure
- Measurement scale for review of work

An MBP approach to operations helps identify unique entities within functional requirements. This identification helps determine where functions overlap and how they are interrelated, thus paving the way for better planning, scheduling, and control. Enterprise-wide project management facilitates a unified view of organizational goals and provides a way for project teams to use information generated by other departments to carry out their functions. The use of project management continues to grow rapidly. The need to develop effective management tools increases with the increasing complexity of new technologies and processes. The life cycle of a new product to be introduced into a competitive market is a good example of a complex process that must be managed with integrative project management approaches. The product will encounter management functions as it goes from one stage to the next. Project management will be needed throughout the design and production stages and will also be needed in developing marketing, transportation, and delivery strategies for the product. When the product finally reaches customers, project management will be needed to integrate its use with those of other products within customer organizations.

The need for a project management approach is established because projects tend to increase in size even if their scope is narrowing. The following four literary laws are applicable to any project environment:

- *Parkinson's law*: Work expands to fill the available time or space.
- *Peter's principle*: People rise to the level of their incompetence.
- *Murphy's law*: Whatever can go wrong will.
- *Badiru's rule*: The grass is always greener where you most need it to be dead.

An integrated systems project management approach can help diminish the adverse impacts of these laws through good project planning, organizing, scheduling, and control.

Integrated Project Implementation

Project management tools can be classified into three major categories:

1. Qualitative tools are the managerial aids that facilitate the interpersonal and organizational processes required for project management.
2. Quantitative tools are analytical techniques that aid in the computational aspects of project management.
3. Computer tools consist of software and hardware that simplify the processes of planning, organizing, scheduling, and controlling a project. Software tools can help in both the qualitative and quantitative analyses needed for project management.

While many books deal with management principles, optimization models, and computer tools, few guidelines exist for the integration of these three areas for project management purposes. In this book, we integrate the three areas as a comprehensive guide to project management. We introduce the *triad approach* to improve the effectiveness of project management with respect to schedule, cost, and performance constraints within the context of systems modeling. Figure 8.4 illustrates the concept, which considers both the management of the project and also the management of all the functions that support the project. It is helpful to have a quantitative model, but

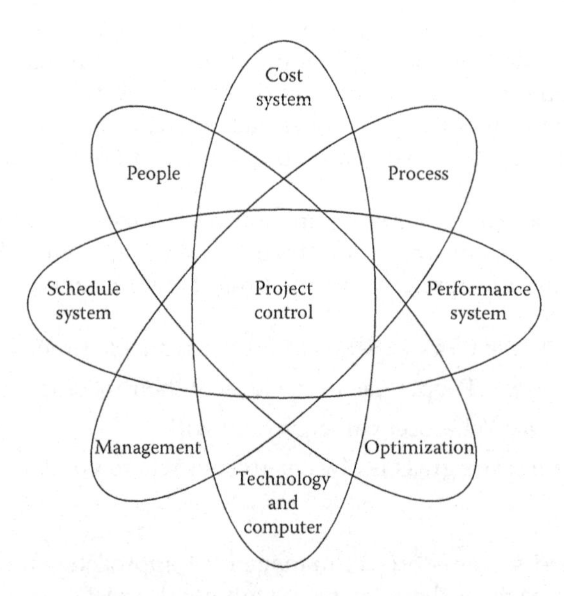

FIGURE 8.4
Systems view encompassing all aspects of a project.

it is far more effective to be able to apply the model to real-world problems in a practical form.

A systems approach helps increase the intersection of the three categories of project management tools and hence improve overall management effectiveness. Crisis should not be the instigator for the use of project management techniques. Project management approaches should be used up front to prevent avoidable problems rather than react to them when they develop. What is worth doing is worth doing well, from the beginning.

Critical Factors for Project Success

The premise of project management is that the critical factors for systems success revolve around people and the personal commitment and dedication of each person. No matter how good a technology is or how enhanced a process may be, the people involved ultimately determine success. This makes it imperative to resolve people issues first in the overall systems approach to project management. Many organizations recognize this, but only a few have been able to actualize the ideals of managing people productively. The execution of operational strategies requires forthrightness, openness, and commitment to get things done. Tangible programs that cater to the needs of people must be implemented. It is essential to provide incentives, encouragement, and empowerment for people to be self-actuating in determining how best to accomplish their job functions.

The first critical factor for systems success is total system management (hardware, software, and people) based on

- Operational effectiveness
- Operational efficiency
- System suitability
- System resilience
- System affordability
- System supportability
- System life cycle cost
- System performance
- System schedule
- System cost

Systems engineering tools, techniques, and processes are essential for project life cycle management to make goals possible in the context of specific, measurable, achievable, realistic, time-bounded (SMART) principles, which are represented as follows:

- Specific: pursue specific and explicit outputs.
- Measurable: design of outputs that can be tracked, measured, and assessed.
- Achievable: make outputs achievable and align them with organizational goals.
- Realistic: pursue only goals that are realistic and result oriented.
- Time bounded: give deadlines to outputs to facilitate accountability.

Project Organization

Project organization specifies how to integrate the functions of the personnel involved in a project. Organizing is usually performed concurrently with project planning. Directing (guiding and supervising project personnel) is an important aspect of project organization and is a crucial aspect of the management function. Directing requires skillful managers who can interact with subordinates effectively via good communication and motivational techniques. A good project manager will facilitate project success by directing his or her staff through proper task assignments toward the project goal.

Employees perform better when they have clearly defined expectations and understand how their job function contributes to the overall goals of the project. Employees should be given some flexibility for self-direction in performing their functions. Individual employee needs and limitations should be recognized by the manager. Directing a project requires motivating, supervising, and delegating skills.

Resource Allocation

Project goals and objectives are accomplished by allocating resources to functional requirements. Resources consist of money, people, equipment, tools, facilities, information, and skills, and are usually in short supply. The people needed for a particular task may be committed to other ongoing projects. A crucial piece of equipment may be under the control of another team.

Project Scheduling

Timeliness is the essence of project management. Scheduling is often the major focus in project management. The main purpose of scheduling is to allocate resources so that the overall project objectives are achieved within a reasonable time span. Project objectives are generally conflicting in nature. For example, minimization of the completion time and minimization of cost are conflicting objectives. That is, one objective is improved at the expense of worsening the other objective. Therefore, project scheduling is a multiple-objective decision-making problem.

In general, scheduling involves the assignment of time periods to specific tasks on the project schedule. Resource availability, time limitations, urgency level, required performance level, precedence requirements, work priorities, technical constraints, and other factors complicate the scheduling process. Thus, the assignment of a time slot to a task does not necessarily ensure that the task will be performed satisfactorily in accordance with the schedule. Consequently, careful trading and monitoring must be performed and controlled throughout project scheduling.

Project Tracking and Reporting

This phase involves checking whether results conform to project plans and performance specifications. Tracking and reporting are prerequisites for project control. A properly organized report of status will help identify deficiencies in the progress of the project and help pinpoint corrective actions.

Project Control

Project control requires that appropriate actions be taken to correct unacceptable deviations from expected performance. Control is actuated through measurement, evaluation, and corrective action. Measurement is the process of measuring the relationship between the planned and actual performance of project objectives. The variables to be measured, the measurement scales, and the measuring approaches should be clearly specified during the planning stage. Corrective actions may involve rescheduling, reallocation of resources, or expedition of task performance. Project control is discussed in detail in Chapter 6 and involves

- Tracking and reporting
- Measurement and evaluation
- Corrective action (plan revision, rescheduling, updating)

Project Termination

Termination is the final stage. The phase-out of a project is as important as its initiation and should be implemented expeditiously. A project should not be allowed to drag on after the expected completion time. A terminal activity should be defined during the planning phase. An example of a terminal activity may be the submission of a final report, the power-on of new equipment, or the signing of a release order. The conclusion of such an activity should be viewed as the completion of the project. Arrangements may be made for follow-up activities to improve or extend the outcome. These follow-up or spin-off projects should be managed as new projects, but with proper input and output relationships within the sequence of the main project.

Project Systems Implementation Outline

While this chapter of project management is aligned with the main tenets of the Project Management Institute (PMI)'s Project Management Body of Knowledge (PMBOK), it follows the traditional project management textbook framework encompassing the broad sequence of categories

Planning → Organizing → Scheduling → Control → Termination

An outline of the functions to be carried out during a project should be made during the planning stage. A model for such an outline is presented in the following section. It may be necessary to rearrange the components of the outline to fit the specific needs of a project.

Planning

1. Specify project background
   a. Define current situation and process

            i. Understand the process
            ii. Identify the important variables
            iii. Quantify the variables
    b. Identify areas for improvement
            i. List and discuss the areas
    c. Study potential strategies for solution
2. Define unique terminologies relevant to the project
    a. Industry-specific terminologies
    b. Company-specific terminologies
    c. Project-specific terminologies
3. Define the project goal and objectives
    a. Write the mission statement
    b. Solicit inputs and ideas from personnel
4. Establish performance standards
    a. Schedule
    b. Performance
    c. Cost
5. Conduct a formal project feasibility study
    a. Determine the impact on cost
    b. Determine the impact on the organization
    c. Determine the project deliverables
6. Secure management support

Organizing

1. Identify the project management team
    a. Specify the project organization structure
            i. Matrix structure
            ii. Formal and informal structures
            iii. Justify the structure
    b. Specify the departments involved and key personnel
            i. Purchasing
            ii. Materials management
            iii. Engineering, design, manufacturing, and so on.

    c. Define project management responsibilities
        i. Select the project manager
        ii. Write the project charter
        iii. Establish project policies and procedures
2. Implement the triple C (Communication, Cooperation, Coordination) model
    a. Communication
        i. Determine the communication interfaces
        ii. Develop the communication matrix
    b. Cooperation
        i. Outline cooperation requirements, policies, and procedures
    c. Coordination
        i. Develop the work breakdown structure
        ii. Assign the task responsibilities
        iii. Develop a responsibility chart

Scheduling (Resource Allocation)

1. Develop the master schedule
    a. Estimate task duration
    b. Identify task precedence requirements
        i. Technical precedence
        ii. Resource-imposed precedence
        iii. Procedural precedence
    c. Use analytical models
        i. Critical path method (CPM)
        ii. Program evaluation and review technique (PERT)
        iii. Gantt chart
        iv. Optimization models

Control (Tracking, Reporting, and Correcting)

1. Establish guidelines for tracking, reporting, and control
    a. Define data requirements
        i. Data categories

        ii. Data characterization
        iii. Measurement scales
    b. Develop data documentation
        i. Data update requirements
        ii. Data quality control
        iii. Establish data security measures
2. Categorize control points
    a. Schedule audit
        i. Activity network and Gantt charts
        ii. Milestones
        iii. Delivery schedule
    b. Performance audit
        i. Employee performance
        ii. Product quality
    c. Cost audit
        i. Cost containment measures
        ii. Percentage completion versus budget depletion
        iii. Identify implementation process
    d. Comparison with targeted schedules
    e. Corrective course of action
        i. Rescheduling
        ii. Reallocation of resources

Termination (Close or Phase-Out)

1. Conduct performance review
2. Develop strategy for follow-up projects
3. Arrange for personnel retention, release, and reassignment

Documentation

1. Document project outcome
2. Submit final report
3. Archive report for future reference

The project team can use the Design for Six Sigma (DFSS) methodology as the framework to develop the project Gantt chart. The team can then also use the tasks to determine the critical path.

Project Plan

The steps for creating the project plan include

1. Identify the tasks to be completed
2. Identify the relationships between the tasks—predecessors and successors
3. Identify the resources necessary for each task
4. Identify the estimated duration and start and end dates
5. Create a network diagram (PERT) or Gantt chart
6. Identify the critical path
7. Monitor tasks

Scope Management

A key focus should remain on the scope outlined in the project charter to avoid scope creep. Occasionally, the scope will need to be expanded or reduced. If this happens, it is important to ensure that these scope changes are documented in the project charter. When changing the project charter, approval must be gained by the team and sponsors, and the modified project charter should be signed off by all team members and stakeholders.

If there are items outside of the project charter scope, these can be captured on an items for resolution (IFR) form, also often referred to as a "parking lot," to manage these items and ensure that they are not forgotten. These may form other improvement projects or can be assigned to teams for resolution outside of the DFSS project. Table 8.1 provides a template for the IFR form.

The purpose of the IFR chart is to document items that should be tracked or resolved. The DFSS team can then use this to prioritize (high, medium, low) these actions and identify the status of each item. It is important to assign owners to allow tracking of each issue as well as the date when the item was opened and when it was resolved. Finally, the resolution should be described to provide history and background.

TABLE 8.1

Items for Resolution Template

No.	Item	Priority	Status	Owner	Open Date	Resolved Date	Resolution

Conclusion

DFSS techniques are used for establishing proper customer requirements for successful Six Sigma business performance. Resource allocation can be performed using tools such as a Gantt chart. The main principle is to assign resources to particular projects broken down by tasks and subtasks with due dates and indicate which tasks need to occur before others.

Resource allocation uses strategic methods to sustain a results-oriented culture. The steering committee of an organization normally allocates the resources. They identify projects, identify Six Sigma Green Belts or Black Belts to use, monitor progress, and establish implementation strategies and policies. The effectiveness of this process must be reviewed to ensure that the steering committee is on the right track.

Resource allocation also consists of ensuring overallocation does not occur. If one particular employee is in every project that the steering committee deems as a priority, that individual is overallocated. The resources should be spread out evenly by the steering committee. Therefore, there should not only be one resource as a key stakeholder; it should be possible to use many individuals in particular projects. In that way, day-to-day activities can be completed normally while projects are ongoing.

Questions

1. Why is project management important in DFSS?
2. What are the key factors in project management?
3. What do the scope requirements determine?

4. What must be achieved within the boundary of the scope requirements?
5. Select one of the reasons projects fail and provide a specific example of how this could occur.
6. What is enterprise-wide project management?
7. What are the characteristics required of each project?
8. Explain the three major categories of project management tools.
9. Why is a systems approach helpful in project management?
10. Give an example of a goal that follows the SMART principle.
11. What is the main purpose of scheduling?
12. In terms of project management, what does measurement refer to?
13. Why is project termination important?
14. What is scope creep?

Technical Design Review: Analyze Phase

In the Analyze phase, the team analyzes the data to identify relationships between key variables, generate new product concepts, and select a new product architecture from the various alternatives. In addition, the team develops multiple conceptual designs and performs a statistical analysis of relevant data to assess the capability of concepts. This involves the development of scorecards to perform a risk assessment.

Technical Design Review

Technical design reviews apply a phase/gate approach to technology development. This is a structured and disciplined use of phases and gates to manage product commercialization. The purpose of a technical design review is to prevent design flaws and mistakes in the next phase of development and to manage obstacles during a development phase.

A phase is a period of time during which specific tools and best practices are used in a disciplined manner to deliver tangible results that fulfill a predetermined set of technology development requirements. A gate is a stopping point within the flow of the phases within a DFSS project.

Two things occur at a gate:

1. A thorough assessment of deliverables from the tools and best practices that were conducted during the previous phase

2. A thorough review of the project management plans for the next phase

A three-tiered system of colors indicates the readiness of a DFSS project to pass through a gate:

1. **Green**: The subsystem has passed the gate criteria with no major problems.
2. **Yellow**: The subsystem has passed some of the gate criteria, but has numerous moderate to minor problems that must be corrected for complete passage through the gate.
3. **Red**: The subsystem has passed some of the gate criteria, but has one or more major problems that preclude the design from passing through the gate.

Checklists can also be used to list tools and best practices required to fulfill a gate deliverable within a phase. The checklist for each phase should be clear so that it is easy to directly relate one or more tools or best practices to each gate deliverable. Product planning can use the items in the checklist in the development of PERT or Gantt charts.

Scorecards are brief summary statements of deliverables from specific applications of tools and best practices. Deliverables should be stated based on the information to be provided, including who is specifically responsible for the deliverable and the due date. This helps the technical design review committee manage risk with keeping the project on schedule.

Gate 2 Readiness: Analyze Phase

- It is essential at this gate to have technology characterized.
- An incomplete investment in this phase will drastically increase costs later.

Assessment of Risks

The following risks should be quantified and evaluated against target values for this phase:

- Quality risks
- Delivery risks
- Cost risks
- Capital risks
- Performance risks
- Volume (sales) risks

9

Gathering the Voice of the Customer

Listening to the customer must become everybody's business. With most competitors moving ever faster, the race will go to those who listen (and respond) most intently.

Tom Peters

Consumer perception of value is critical for any product or service provider, regardless of the industry. Consumers seek value in products, but not all consumers view value the same. This naturally leads to market segmentation. Product and service providers, therefore, must address the needs of various market niches to enhance their overall market penetration and share. Profitability for organizations depends on quality, cost, and the timely delivery of their products and services. The most important driver is quality because it can be used to drive down cost for the producer and the consumer, improve productivity, and facilitate fast product–development cycles and short delivery times. Accordingly, the goal of the producer is to deliver a product or service that is reliable and durable and has long-term resale value while creating consumer enthusiasm.

Consumers desire a high level of performance, capability, appeal, and style relative to an associated cost; therefore, companies must strive to maximize product function relative to its cost. In order to accomplish this, companies must understand the voice of the consumer (VOC) and the consumer needs.

The VOC represents the desires and requirements of the customer at all levels, translated into real terms for consideration in the development of new products, services, and daily business conduct. The VOC is a method for listening to and understanding the customer needs. It is important to have constant contact with the customer to understand their changing needs.

VOC in Product Development

It is critical to involve the customer(s) early in defining product, process, or service requirements. The requirements must be based on customer needs to ensure that the final product is marketable. The requirements are then used to determine the appropriate specifications necessary to meet the high-level requirements.

Customers/Stakeholders

At the beginning of the Design for Six Sigma (DFSS) project, it is essential for the team to determine the customers and stakeholders. Customers typically include:

- Peers
- Associates in the reporting line
- Supervisors
- Suppliers
- Other business areas
- Other groups in the organization
- External customers

The team must identify primary and secondary stakeholders and anyone affected by the project. Once the stakeholders are identified, the team needs to ensure that they understand the stakeholders' attitudes toward change and any potential reasons for resistance. As part of this effort, the team should develop activities, plans, and actions to overcome the resistance and barriers to change. To do this, the team should determine

- How and when each stakeholder group should participate in the change
- How to gain buy-in during the planning phase

Table 9.1 provides a matrix to identify stakeholder groups, their role in the project, and their impact and/or concerns.

TABLE 9.1

Customer/Stakeholder Matrix

Stakeholder/ Customer	Role Description	Impact/Concern	+/–

Voice of the Customer

There are several ways to gather the VOC. It is important to gather the VOC using several methods to capture:

- Likely satisfiers: Needs known to and voiced by customers
- Likely dissatisfiers: Needs known to but not voiced by customers
- Likely delighters: Needs that customers are unaware of

When gathering the VOC, there are several aspects to keep in mind to ensure that you are "listening" to the customer.

- You are not the customer: Do not use your instincts
- View the product from the perspective of what the customer sees and experiences
- 90%–96% of unhappy customers do not complain: Determine the most appropriate way to get them to discuss their experience and gain their feedback

The various methods for gathering the VOC for these categories are provided in Table 9.2.

Surveys provide a method for gathering sample data. Surveys can be distributed via mail, e-mail, or phone. Properly designed surveys gather demographic information and questions regarding the customers' requirements. Mail and e-mail surveys can be widely distributed to gather information from a broad range of current customers, potential customers, and past customers. E-mail surveys, in particular, are relatively inexpensive. The downside, however, is that response rates are relatively low—typically 5%. In addition, unless there is an incentive, participants may be reluctant to respond unless they had a strong positive or a strong negative experience.

Focus groups consist of 3–12 participants. A narrative protocol is used to guide the discussion using open-ended questions. This allows the interviewers to explore specific topics regarding the product or service

TABLE 9.2

Data Collection Methods

Likely Satisfiers	Likely Dissatifiers	Likely Delighters
• Surveys: Mail, e-mail, and phone • Interviews: Face-to-face, and phone • Focus groups • Market research • Competitor marketing and ads	• One-on-one interviews • Industry standards and regulatory requirements • Internal customer complaint data • Publicly available recall data	• Structured and planned focus groups • Observe customers • Painstorming • Innovation

TABLE 9.3
VOC Data Collection Plan

Critical to Satisfaction (CTS)	Metric	Data Collection Method (Survey, Interview, etc.)	Analysis Plan (Statistical Test, Graphical Analysis, etc.)	Sampling Plan (Size, Frequency, etc.)	Sampling Details (Who, Where, When, How)

being redesigned. Focus groups enable further information such as facial expressions and follow-on clarification questions that cannot be captured in surveys. In addition, comments by participants can spark further ideas.

One-on-one interviews typically last 30–60 minutes. This methodology allows the team to gather significant information about the customers' wants and needs and their opinions on competitor products. A narrative protocol is also used for one-on-one interviews to guide the discussion. However, since only one participant is interviewed at a time, this is a time-consuming process.

Another method for gathering the VOC is the use of satisfaction and complaint cards. Complaints are gold! They provide very direct verbatims from the customer on what they dislike and on what should be improved. Conversely, satisfaction cards provide data on what is working well. Since this information is typically already captured as part of normal business operations, it does not require additional resources to gather the data, only to analyze the data for patterns and trends. Dissatisfaction data can come from complaints, refunds, returns, warranties, replacements, claims, recalls, and litigation, among others.

Each methodology for gathering the VOC has its benefits and constraints. A well-rounded approach typically entails multiple methods to gather the most information. Once you have determined the approach for gathering the VOC, document it in a data collection plan form as shown in Table 9.3.

Critical to Satisfaction

Critical to satisfaction (CTS) characteristics are the elements within a process that impact the process output. Therefore, these are the elements of the process that should be managed and measured because they have a direct effect on the performance. In other words, these are the process characteristics that

TABLE 9.4

CTS Characteristics

Voice of the Customer (Verbatim)	Why? (After Clarification)	Target	Tolerance
"Ordering from your company is difficult."	Shipments take too long to receive.	Five business days' lead time	+/− one day

are critical to what the customer perceives as quality. The steps to develop the CTS characteristics include

1. Gather the VOC data using the methods described in Table 9.2.
2. Determine the relevant statements from the customer transcripts. It is important here to use the customer verbatims: Exactly what the customer said so that we are not putting any of our own bias into translating what the customer meant. These statements should then be placed on sticky notes.
3. Sort the ideas and find the common themes using affinity diagrams or tree diagrams. It is important to focus on the key statements that relate to whether or not a customer would or would not buy the product.
4. Use the themes to determine why the customer feels this way. Conduct follow-up focus groups and/or customer interviews to clarify statements.
5. Conduct additional interviews and/or focus groups to determine quantifiable targets and specifications.

As the CTS information is gathered, it should be documented. An example template is provided in Table 9.4. As the team develops the CTS characteristics, it is important to keep several aspects in mind. The CTS should be specific; otherwise it will lead to confusion. The CTS should also be measureable to determine if it is within the specification limits.

Critical to Quality

Critical to quality (CTQs) are characteristics that are important to the customer. They come from the VOC. CTQs are measureable and quantifiable metrics that come from the VOC. An affinity diagram is an organizational tool for VOCs.

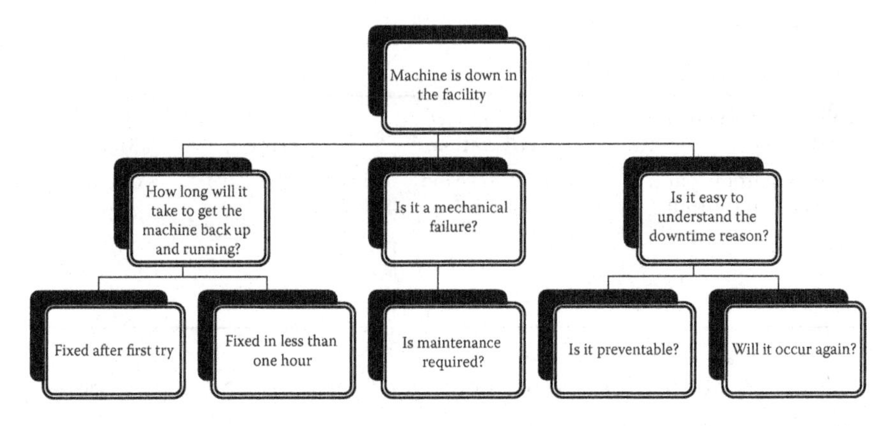

FIGURE 9.1
CTQ tree.

CTQs are critical to communication because we need to understand the critical aspects that matter most to the customer.

Utilizing a VOC for manufacturing internally is a good way to understand processes that the employees know a great deal about. Therefore, the production worker(s) are the customers and the questions are given to them. A tree format helps with the visualization of CTQs as shown in Figure 9.1.

Conclusion

Gathering the VOC is a critical part of DFSS to ensure the desires and requirements of the customer are understood at all levels. The DFSS team can then translate the VOC into the development of new products, services, processes, and daily business conduct. The VOC cannot be gathered just at the beginning of the product development process because this only captures one point in time and customers' needs and desires change. It is important to have constant contact with the customer to understand their changing needs. The next chapter discusses quality function deployment, which is a methodology for translating customer needs into design requirements to ensure customer satisfaction.

Questions

1. When should customer involvement begin in DFSS?
2. What are customer requirements that are specifically requested, which if present the customer is satisfied and if absent the customer is dissatisfied, known as?

3. Give an example of a likely dissatisfier.
4. What are the benefits of surveys? What are the disadvantages of surveys?
5. What are the benefits of focus groups? What are the disadvantages of focus groups?
6. What are the benefits of one-on-one interviews? What are the disadvantages of one-on-one interviews?
7. Why should you use multiple methods to gather the VOC?
8. What are critical to satisfaction characteristics? Give an example.
9. What is a verbatim? Why is it important to use?
10. What method is used to process VOC data into rational groups based on common themes?

10

Quality Function Deployment

Because it is customers who must buy the product and who must be satisfied with it, the product must be developed with their needs and wants as the principal inputs to the new product development process. When this is not the case, the new product introduction is often disappointing.

Ronald G. Day

Translating the voice of the consumer (VOC) into product characteristics is vital for a company to remain competitive. Quality function deployment (QFD) is a systematic approach for translating consumer requirements into appropriate company requirements at each stage from research and product development to engineering and manufacturing to marketing/sales and distribution. QFD makes use of the voice of the consumer throughout the process. Design for Six Sigma (DFSS) addresses customer needs and product value using QFD.

Kano Model

The Kano Model was developed by Noriaki Kano in the 1980s. The Kano Model is a methodology that defines three types of quality requirements and shows how achieving these requirements affects customer satisfaction. The three quality requirements are one-dimensional, must-be or basic requirements, and attractive requirements, as shown in Figure 10.1. The Kano Model helps identify critical-to-quality (CTQs) characteristics that add incremental value versus those that are simply requirements where having more is not necessarily better.

The Kano Model engages customers by understanding the product attributes that are most important to customers. The purpose of the tool is to support product specifications that are made by the customer and promote discussion while engaging team members. The model differentiates features of products rather than customer needs by understanding necessities and items that are not required. Kano also produced a methodology for mapping consumer responses with questionnaires that focused on attractive qualities through reverse qualities.

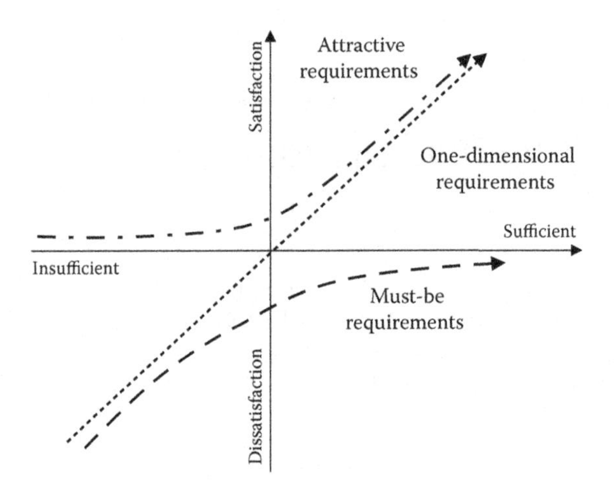

FIGURE 10.1
Kano Model.

One-dimensional requirements are specifically requested items. These are often stated wants from a survey, focus group, or interview. If these are present, the customer is satisfied because they received what they requested. However, if these requirements are absent, then the customer is dissatisfied. For example, this could be ordering a shirt in a medium or requesting a special order at a restaurant.

Must-be or basic requirements are those requirements that are not specifically requested, but assumed present. If these are present, the customer is neither satisfied nor dissatisfied. However, if they are absent, the customer is very dissatisfied. For example, clean silverware in a restaurant is a must-be requirement.

Attractive requirements are unknown to the customer. Therefore, they are the most difficult to define and develop. If these are present, the customer is very satisfied as this would be an unexpected surprise. These are also commonly referred to as delighters. However, if these are absent, the customer is neither satisfied nor dissatisfied because the customer was not aware of them.

A Kano table is provided in Table 10.1.

The Kano Model is important to use for sustainability because it differentiates which aspects or requirements must be accomplished to protect our environment and which aspects can gradually be improved upon.

Quality Function Deployment

Dr. Yoji Akao developed QFD in 1966 in Japan. A combination of quality assurance and quality control led to value engineering analyses. The methods

TABLE 10.1
Kano Model Table

Answers to a Positively Formulated Question	Answers to a Negatively Formulated Question			
	I Like That	That's Normal	I Don't Care	I Don't Like That
I like that	—	Delighter	Delighter	Satisfier
That's normal	—	—	—	Dissatisfier
I don't care	—	—	—	Dissatisfier
I don't like that	—	—	—	—

for QFD are simply to utilize consumer demands in designing quality functions and methods to achieve quality in subsystems and specific elements of processes. The basis for QFD is to take customer requirements from the voice of the customer and relay them into engineering terms to develop products or services. Graphs and matrices are utilized for QFD. A house-type matrix is compiled to ensure that the customer needs are being met in the transformation of the processes or services designed. QFD is a collection of matrices that are used to facilitate group discussions and decision-making. The QFD matrices are constructed using affinity diagrams, brainstorming, decision matrices, and tree diagrams.

As a structured team approach, QFD can lead to a better understanding of customer requirements, which, in turn, leads to increased customer satisfaction and better teamwork. Also, understanding and integrating the VOC into the design reduces the time to market and development costs. QFD also enables a structured integration of competitive benchmarking (covered in Chapter 7) in the design process. QFD also increases the team's ability to create innovative design solutions that provide enhanced capability and greater customer satisfaction. Finally, QFD provides documentation of the key design solutions.

There are seven key steps in creating the house of quality (HOQ).

1. Determine what the customer wants. These are the "whats."
2. Determine how well the organization meets the customer requirements compared with the competition. This is the competitive assessment.
3. Determine where the focus should be to maximize return. This is based on the competitive assessment.
4. Develop methods to measure or control the product or process to ensure customer requirements are met. These are the "hows."

5. Evaluate the proposed design requirements (hows) against the customer requirements (whats). This is performed in the relationship matrix.
6. Evaluate the design trade-offs. This is performed in a correlation matrix or "roof" of the HOQ.
7. Determine the key design requirements that should be focused on. This is in the "basement" of the HOQ and is commonly referred to as the "how much."

A general overview of the HOQ is provided in Figure 10.2.

The QFD house is a simple matrix where the legend is used to understand quality characteristics, customer requirements, and completion. QFD is a structured methodology to identify and translate customer needs (wants) into technical requirements, measurable features, and characteristics:

- Marketing and sales
- Research and product development
- Engineering and manufacturing
- Distribution and services

QFD ensures that an orderly process is conducted for determining CTQs while utilizing a common-sense approach. It is important to understand what the customer wants by very intuitive listening.

It is also important to understand when it is appropriate to utilize QFD. The following are key areas for QFD implementation:

- Complex product development initiatives
    - Communications flowdown difficulties
    - Expectations get lost

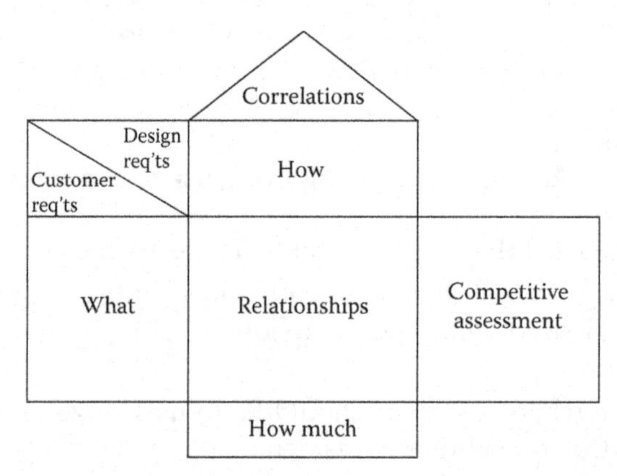

FIGURE 10.2
House of quality.

- New product initiatives/inventions
    - Lack of structure or logic to the allocation of development resources
- Large, complex, or global teams
    - Lack of efficient and/or effective processes
    - Teamwork issues
- Extended product development times
    - Excessive redesign
    - Changing teams
    - Problem solving or firefighting

QFD works as a basic building block with a structured flowdown process. All CTQs are identified at one level, and the related CTQs are built in the next level followed by the "hows." The "hows" at the next level become the "whats" at the next level. QFD follows a four-phase approach:

- Phase I: Product Planning Phase
    - Customer requirements are identified and used to prioritize design requirements
    - HOQ is constructed
- Phase II: Part Deployment Phase
    - The prioritized design requirements are used to identify key part characteristics
- Phase III: Process Planning Phase
    - Key parts are used to systematically identify those process operations that are critical to ensuring customer requirements
- Phase IV: Production Planning Phase
    - Specific production requirements are identified to ensure that key processes from Phase III are controlled and maintained

The HOQ is the initial QFD matrix. It is where the customer requirements are analyzed using affinity and tree diagrams. This house is shown in Figure 10.3. The graphical representation of the flow to creat the HOQ is shown in Figure 10.4.

The customer requirements represent the voice of the customer or the "whats." The whats in the HOQ consist of:

- What does the customer want
- Often called customer needs
- High-level CTQs
- The Y's

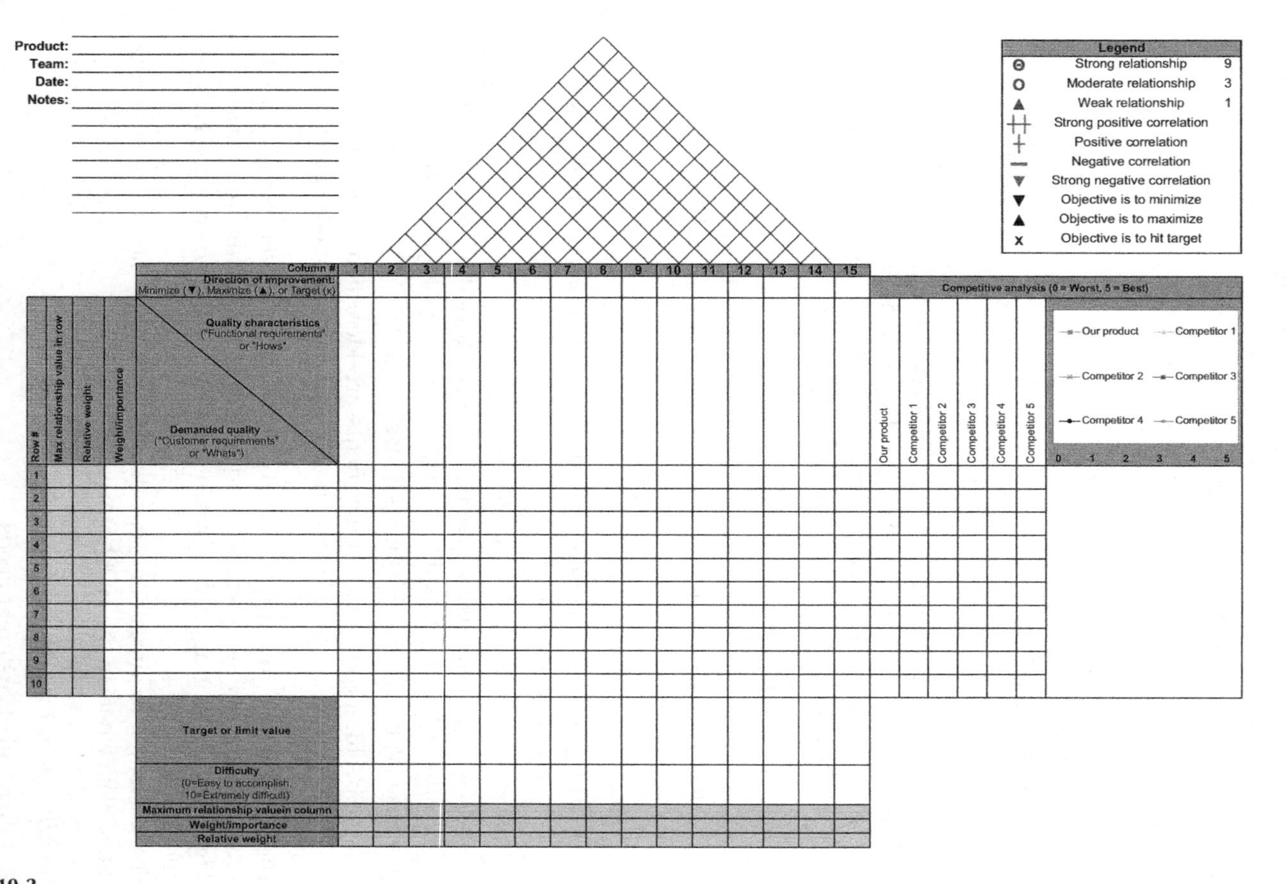

FIGURE 10.3
Quality function deployment house.

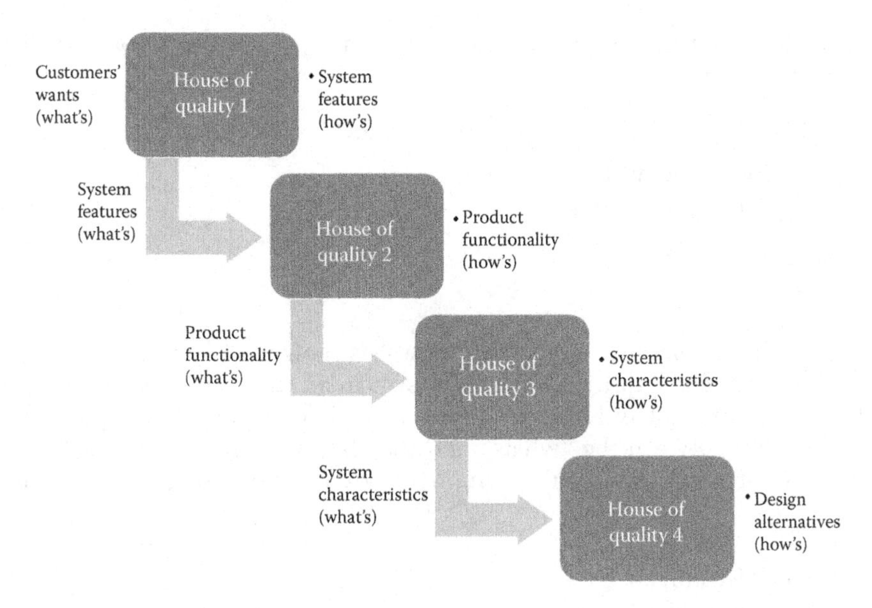

FIGURE 10.4
House of quality graphical representation.

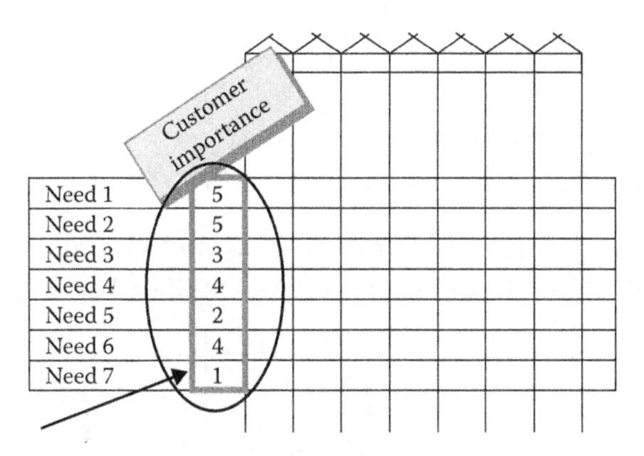

FIGURE 10.5
Voice of the customer ranking.

These are conducted using the voice of the customer via brainstorming. The voice of the customer is validated, prioritized, and benchmarked using direct feedback from the customer. The customer importance is ranked in order of their needs—how important are the "whats" to the customer demonstrated in Figure 10.5.

The design requirements are the "hows." These are brainstormed and analyzed similarly to the "whats." The design requirements represent the aspects of the product or service being designed that should be measured to ensure that all of the customer requirements are met. The "hows" in the HOQ consist of

- How do you satisfy the customer "whats"
- Actionable items to meet needs
- Translation for action
- Quantifiable
- X's

A graphical demonstration of the "hows" is shown in Figure 10.6.

Each customer requirement is then systematically compared with each design requirement and the strength of each relationship is determined. The relationship between the "whats" and the "hows" is the key element. The influence of each "how" on the "what" can be correlated using the following:

- H=Strong (9)
- M=Medium (3)
- L=Weak (1)
- Can leave blank: No influence

Each row is a transfer function where

- Y=f(X's)
- Low fidelity, qualitative
- Need 1=f (How 1, How 2, How 5, How 7)

The relationship matrix should then be analyzed to determine if there are any potential issues with the proposed design. The relationship matrix

How's

		How 1	How 2					How 7
Need 1	5							
Need 2	5							
Need 3	3							
Need 4	4							
Need 5	2							
Need 6	4							
Need 7	1							

FIGURE 10.6
Satisfy the customer's needs via "hows."

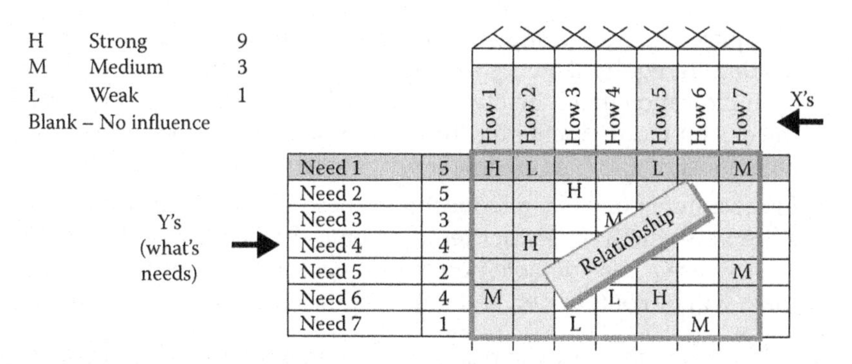

FIGURE 10.7
Relationship for the house of quality.

is shown in Figure 10.7. There are seven relationships that should be evaluated including

1. Blank columns: If a column is blank, then it is not addressing any of the customer requirements. This can be an indication of an unnecessary design requirement or an attractive requirement factor based on the Kano analysis.
2. Blank rows: If a row is blank, then the customer requirement is not being addressed by any of the design requirements. This is an indication that a customer requirement was overlooked in the design. With respect to the Kano Model, this would be a one-dimensional requirement because the customer specifically asked for it. Therefore, the customer will be dissatisfied if it is absent.
3. Rows with no strong relationships: This is similar to a missed customer requirement since there will not be a strong relationship with a design requirement. The customer will not be able to clearly see and feel the value of their request.
4. Identically weighted rows: Typically, customer requirements will be addressed through different design requirements. Therefore, two rows should not be weighted identically. This is an indication of a possible misunderstand of customer requirements since both customer requirements would be addressed through the same design requirements.
5. Complete or nearly complete row: If a row is complete or nearly complete, this is an indication that the customer requirement will involve cost, reliability, or safety problems because the majority or all of the design requirements are addressing the customer requirement. If the customer requirement is no longer needed, this affects the entire design. Since most products or services are designed to meet the demands of several market segments, this would impact the design significantly.

6. Complete or nearly complete column: Similar to evaluation Point 5, if a design requirement has a complete or nearly complete column, it is an indication that it involves cost, reliability, or safety problems. If the design requirement fails, then the product or service will fail to meet most or all of the customer requirements.
7. Large number of weak relationships: There should be clear relationships between the customer requirements and design requirements. If there are a considerable number of weak relationships, the decisions will be confounded and not meet the customer's expectations.

Calculations are then used to prioritize customer requirements and design requirements. This enables the team to focus on the key requirements that have the greatest impact on the customer. The technical importances or key elements should be ranked via sorting on the technical importance values. It is important to identify the critical few and where the focus should lie as shown in Figure 10.8.

Finally, the following steps need to be taken for accuracy:

- Interpreting technical importance
  - If focused on as a group allows the maximum opportunity of meeting all of the customer needs simultaneously
  - The absolute value of the technical importance has no significance other than ranking the "hows" for interpretation
  - High return on investment (ROI) elements could be anywhere
- Completeness: Key elements
  - Is the transfer function complete?
  - Are all of the "hows" captured?

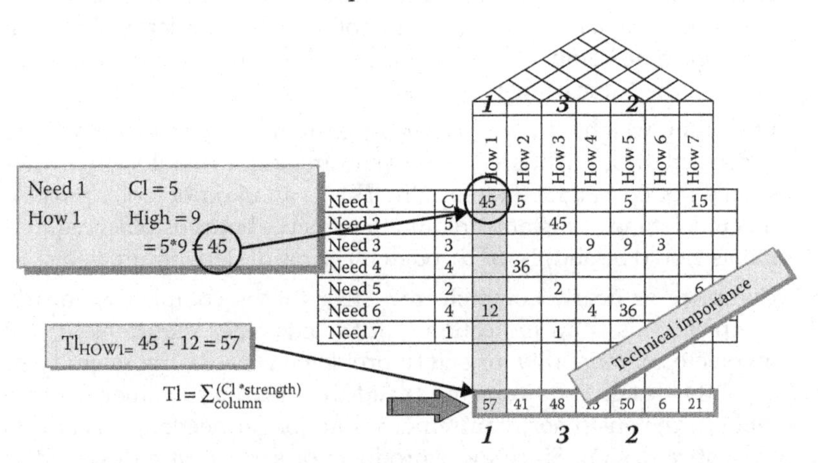

FIGURE 10.8
Technical importance key elements.

- Is a "what" really a "how"?
- Identify point solutions

A completeness check shows if the "hows" have been captured (Figure 10.9). The key elements for completeness include looking for red flags, as shown in Figure 10.10.

When completing and interpreting the QFD, the following focus should be taken:

- Focus on the transfer function nature of the row
  - Evaluate high, middle, low, and blanks as a function $Y = f(x)$

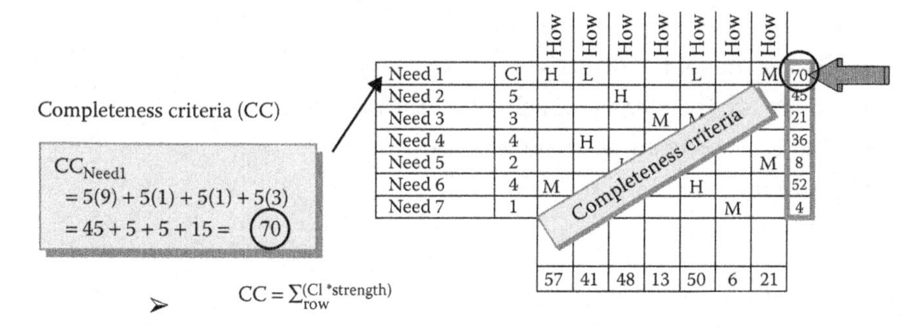

FIGURE 10.9
Completeness check.

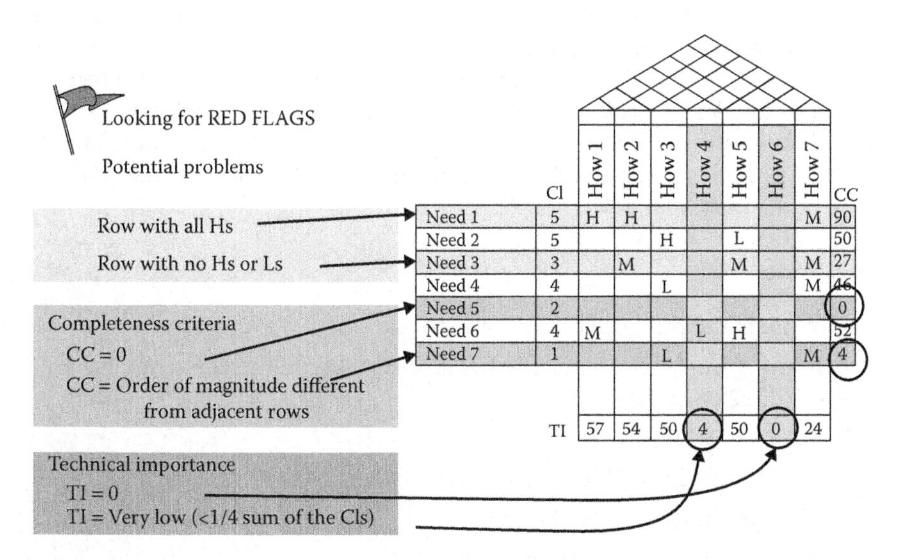

FIGURE 10.10
Looking for red flags.

- Technical importance
  - No direct relationships
  - Interpreted by comparing magnitudes
- Ensure QFD is updated to avoid expiration
  - Review at least quarterly, more often on small projects

The final items to remember for QFDs are the following:

- The process may look simple, but requires effort.
- Many entries look obvious after they are written down.
- If there are *no* "tough spots" the first time: *It is probably not being done right*!
- Focus on the end-user customer.
- Charts are not the objective.
- Charts are the means for achieving the objective.
- Find reasons to succeed, not excuses for failure.
- Remember to follow up afterward.

Conclusion

QFD provides a systematic approach for translating the voice of the customer into product characteristics. QFD ensures the voice of the customer is heard throughout research and product development to engineering and manufacturing to marketing/sales and distribution. Design for Six Sigma addresses customer need and product value using QFD to design products, services, and processes that meet or exceed customer expectations. The next chapter on TRIZ provides a methodology to develop innovative designs based on the voice of the customer.

Questions

1. What is the purpose of the Kano Model?
2. In the Kano Model, what are the three quality requirements? Give an example of each.
3. What is QFD?
4. What is the initial QFD matrix known as?
5. What are the benefits of QFD?
6. What do the "whats" refer to in QFD?
7. What does a blank row indicate in QFD?

11

TRIZ

I believe in being an innovator.

Walt Disney

TRIZ (Theoria Resheneyva Isobretatelskehuh Zadach) is a Russian acronym for the "theory of inventive problem solving." The term is pronounced as "trēz" or "trees." The methodology is also sometimes called TIPS because of the English translation. It is an analytical methodology that was developed by Genrich S. Altshuller, a Russian inventor and naval officer, in 1946. The methodology was developed through the analysis of over three million patents. TRIZ is a theory based on the concept that there are universal principles that drive invention and creativity.

TRIZ is considered a "left brain engineering" method for developing solutions. It is a structured innovation method for rapidly developing technical and nontechnical concepts and solutions. It focuses on the study of patterns of problems and solutions. TRIZ is based on the concept that most underlying root problems have already been solved—either in a different industry, in an unrelated situation, or using a different technology. One of the key ideas in TRIZ is that technical systems evolve for the better. This is accomplished by overcoming contradictions in the technical system though minimizing the introduction of resources. Therefore, TRIZ is useful in Design for Six Sigma (DFSS) for the following:

- Creating new design solutions
- Resolving design contradictions
- Increasing design options

TRIZ Methodology

Altshuller studied global patents across different fields to determine how innovation had taken place and to codify the methodology. The analysis showed consistent patterns of invention across these patents and consistent technological evolution across domains. His hypothesis was that there are universal principles for creativity that lead to creative innovations, which, in turn, advance technology. In his study, he discovered that while each innovation patent solved a technical problem, there were common solutions across domains. The general concept behind the methodology is that a problem (or something very similar) has already been solved somewhere. Therefore, the necessary creativity involves finding that previous solution and adapting it to the problem at hand. Altshuller believed that if these principles were codified, they could then be taught to make the creative process more predictable.

TRIZ offers several benefits, the most notable being that it helps solve problems. It also provides a systematic, structured approach that allows teams to think clearly and be creative and innovative. TRIZ also helps to improve systems by increasing the ideality of systems through lowering costs and increasing benefits. This also helps organizations use their resources more effectively by finding inexpensive and quick solutions to difficult problems. The TRIZ ideality/ideal design for a system is centered on the premise that an ideal system performs the desired function without the physical existence of any system. An ideal system occupies no space, takes no time, and does not require labor or maintenance. It is a robust system, which means that the system is insensitive to noise factors. The measure is the sum of the benefits from the system divided by the sum of the costs and harms, as shown in Equation 11.1. The benefits refer to the useful function or desired outcome. The costs are the direct costs and costs to society (similar to Taguchi's philosophy with the quadratic loss function). The harms are the failure modes, harmful side effects, and other undesired outcomes.

$$\text{Degree of ideality} = \frac{\text{Benefits}\left(\text{functionality}\right)}{\text{Cost} + \text{harmful side effects}} \tag{11.1}$$

The goal is to focus on the solution rather than focusing on the problems. This enables teams to determine how best to use their resources to achieve the solution.

Nine Windows

Nine Windows is a commonly used tool within TRIZ. The tool is based around the concept that we normally view the world through one window. This tool forces us to view and evaluate the world through nine windows by taking into account the past, present, and future in combination with the system, subsystem, and supersystem levels. It is presented as a 3×3 matrix, as shown in Figure 11.1. By looking at the past, present, and future, we can gain a historical view and the context of the problem at hand. Further, by looking at the problem from a system, the system and the environment (supersystem),

	Past	Present	Future
Supersystem			
System			
Subsystem			

FIGURE 11.1
Nine windows matrix.

and the subsystem, we can understand the overall system and its structure/relationships.

If the problem is not fully resolved, then several TRIZ database tools can be applied. These include the contradiction matrix and separation principles.

TRIZ Methodology

TRIZ provides a reliable, repeatable, and systematic method for innovation. The TRIZ method involves four key steps, as shown in Figure 11.2. Within this methodology, it is important to understand the problem. The focus in the

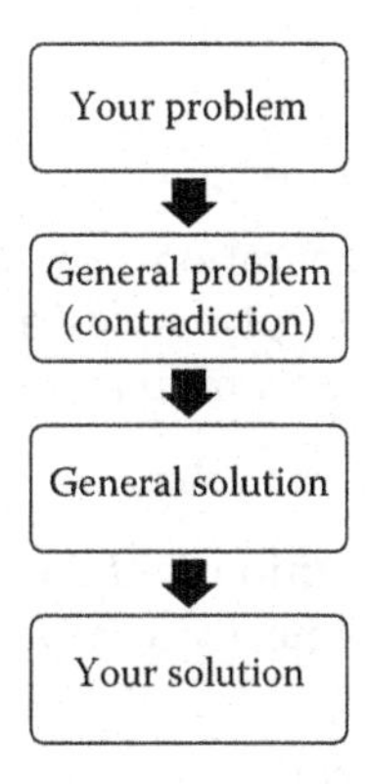

FIGURE 11.2
TRIZ method.

methodology, as shown in Equation 11.1, is on the benefits and not the "how." The team should envision the ideal outcome through consensus. Using this information, the team can determine the benefits and future state and then visualize the solution. This enables the team to search for the appropriate resources to achieve these benefits. Finally, the team can identify contradictions using TRIZ for further improvement.

Contradictions

Altshuller found that inventive problems typically have inherent contradictions. A key concept in TRIZ is that fundamental contradictions are the root cause of many problems. Within TRIZ, there are two key categories of contradictions: technical contradictions and physical contradictions.

Technical Contradictions

Technical contradictions are the engineering trade-offs that must be made during design. In other words, this means that as something gets better, something else gets worse. For example, the product might need to be made stronger, but that would also result in the product being heavier.

These are solved through the 39 elimination principles.

Physical Contradictions

Physical contradictions are inherent contradictions that occur when a system has contradictory requirements. For example, a new cellular phone should have many powerful features; however, it should also be easy to use.

These are solved using four separation principles that address supersystems, subsystems, and separation of time and space.

Separation Principle

Within TRIZ, there are four separation principles that can be used to generate new ideas. The separation principles are typically used early in the process because they can aid the team in proposing a solution quickly. These principles include

- Separation in time: Determine whether the actions, parts, or system can be separated in time. Can they happen before or after each other?
- Separation in space: Determine whether the conflict can be resolved by physically moving the actions, parts, or system. Conversely, if

there already is a separation in space, determine whether removing the separation in space will resolve the conflict.

- Separation between components and the system: Determine whether the actions, parts, or system can be separated into smaller parts. Conversely, determine whether they can be combined.
- Separation on condition: Determine whether the actions, parts, or systems can be treated differently based on the condition (internal or external).

Contradiction Matrix

The contradiction matrix is one of the main tools within TRIZ for solution generation. It is typically considered the starting point after the separation principles. The contradiction matrix is used when two features are determined to be in a trade-off situation. The goal of TRIZ is to eliminate these contradictions to allow both features to be included without contradiction. This is where innovation comes into the process. Designers would normally be forced to compromise between the two extremes of the contradictory features.

The contradiction matrix is a 39×39 matrix. Each side includes the 39 parameters, and each entry includes one to four inventive principles that should be considered (Figure 11.3 through 11.5).

The 40 Principles of Invention

In addition, while innovative patents are developed to solve a technical problem in one domain, there are common solutions across domains. These common solutions were reduced to 40 principles. The 40 principles of TRIZ are basic engineering parameters. These parameters include aspects such as height and weight. The theory states that the parameters that are causing the problem should be analyzed against the parameters it is conflicting with to develop engineering solutions. The 40 principles are shown in Figure 11.6.

TRIZ and DFSS

TRIZ is particularly useful in DFSS when the team reaches a point where they are unclear what steps to take for resolution. Progress can seem blocked, and the path forward to reach the organization's goals may be unclear. This is when the team needs to be creative to develop innovative solutions. Divergence tools such as brainstorming exist to help develop a large list of ideas; however, this is reliant on the experience, intuition, and knowledge of the team. Therefore, there is a risk of missing possible solutions. TRIZ is a methodology that can minimize the unpredictability and unrepeatability

	Improving feature ↓ / Worsening feature →	Weight of moving object	Weight of stationary object	Length of moving object	Length of stationary object	Area of moving object	Area of stationary object	Volume of moving object	Volume of stationary object	Speed	Force (intensity)	Stress or pressure	Shape	Stability of the object's composition
		1	2	3	4	5	6	7	8	9	10	11	12	13
1	Weight of moving object	+		15, 8, 29,34		29, 17, 38, 34		29, 2, 40, 28		2, 8, 15, 38	8, 10, 18, 37	10, 36, 37, 40	10, 14, 35, 40	1, 35, 19, 39
2	Weight of stationary object		+		10, 1, 29, 35		35, 30, 13, 2		5, 35, 14, 2		8, 10, 19, 35	13, 29, 10, 18	13, 10, 29, 14	26, 39, 1, 40
3	Length of moving object	8, 15, 29, 34		+		15, 17, 4		7, 17, 4, 35		13, 4, 8	17, 10, 4	1, 8, 35	1, 8, 10, 29	1, 8, 15, 34
4	Length of stationary object		35, 28, 40, 29		+		17, 7, 10, 40		35, 8, 2,14		28, 10	1, 14, 35	13, 14, 15, 7	39, 37, 35
5	Area of moving object	2, 17, 29, 4		14, 15, 18, 4		+		7, 14, 17, 4		29, 30, 4, 34	19, 30, 35, 2	10, 15, 36, 28	5, 34, 29, 4	11, 2, 13, 39
6	Area of stationary object		30, 2, 14, 18		26, 7, 9, 39		+				1, 18, 35, 36	10, 15, 36, 37		2, 38
7	Volume of moving object	2, 26, 29, 40		1, 7, 4, 35		1, 7, 4, 17		+		29, 4, 38, 34	15, 35, 36, 37	6, 35, 36, 37	1, 15, 29, 4	28, 10, 1, 39
8	Volume of stationary object		35, 10, 19, 14	19, 14	35, 8, 2, 14				+		2, 18, 37	24, 35	7, 2, 35	34, 28, 35, 40
9	Speed	2, 28, 13, 38		13, 14, 8		29, 30, 34		7, 29, 34		+	13, 28, 15, 19	6, 18, 38, 40	35, 15, 18, 34	28, 33, 1, 18
10	Force (intensity)	8, 1, 37, 18	18, 13, 1, 28	17, 19, 9, 36	28, 10	19, 10, 15	1, 18, 36, 37	15, 9, 12, 37	2, 36, 18, 37	13, 28, 15, 12	+	18, 21, 11	10, 35, 40, 34	35, 10, 21
11	Stress or pressure	10, 36, 37, 40	13, 29, 10, 18	35, 10, 36	35, 1, 14, 16	10, 15, 36, 28	10, 15, 36, 37	6, 35, 10	35, 24	6, 35, 36	36, 35, 21	+	35, 4, 15, 10	35, 33, 2, 40
12	Shape	8, 10, 29, 40	15, 10, 26, 3	29, 34, 5, 4	13, 14, 10, 7	5, 34, 4, 10		14, 4, 15, 22	7, 2, 35	35, 15, 34, 18	35, 10, 37, 40	34, 15, 10, 14	+	33, 1, 18, 4
13	Stability of the object's composition	21, 35, 2, 39	26, 39, 1, 40	13, 15, 1, 28	37	2, 11, 13	39	28, 10, 19, 39	34, 28, 35, 40	33, 15, 28, 18	10, 35, 21, 16	2, 35, 40	22, 1, 18, 4	+
14	Strength	1, 8, 40, 15	40, 26, 27, 1	1, 15, 8, 35	15, 14, 28, 26	3, 34, 40, 29	9, 40, 28	10, 15, 14, 7	9, 14, 17, 15	8, 13, 26, 14	10, 18, 3, 14	10, 3, 18, 40	10, 30, 35, 40	13, 17, 35
15	Duration of action of moving object	19, 5, 34, 31		2, 19, 9		3, 17, 19		10, 2, 19, 30		3, 35, 5	19, 2, 16	19, 3, 27	14, 26, 28, 25	13, 3, 35
16	Duration of action by stationary object		6, 27, 19, 16		1, 40, 35				35, 34, 38					39, 3, 35, 23
17	Temperature	36,22, 6, 38	22, 35, 32	15, 19, 9	15, 19, 9	3, 35, 39, 18	35, 38	34, 39, 40, 18	35, 6, 4	2, 28, 36, 30	35, 10, 3, 21	35, 39, 19, 2	14, 22, 19, 32	1, 35, 32
18	Illumination intensity	19, 1, 32	2, 35, 32	19, 32, 16		19, 32, 26		2, 13, 10		10, 13, 19	26, 19, 6		32, 30	32, 3, 27
19	Use of energy by moving object	12,18,28 ,31		12, 28		15, 19, 25		35, 13, 18		8, 35, 35	16, 26, 21, 2	23, 14, 25	12, 2, 29	19, 13, 17, 24
20	Use of energy by stationary object		19, 9, 6, 27								36, 37			27, 4, 29, 18
21	Power	8, 36, 38, 31	19, 26, 17, 27	1, 10, 35, 37		19, 38	17, 32, 13, 38	35, 6, 38	30, 6, 25	15, 35, 2	26, 2, 36, 35	22, 10, 35	29, 14, 2, 40	35, 32, 15, 31
22	Loss of energy	15, 6, 19, 28	19, 6, 18, 9	7, 2, 6, 13	6, 38, 7	15, 26, 17, 30	17, 7, 30, 18	7, 18, 23	7	16, 35, 38	36, 38			14, 2, 39, 6
23	Loss of substance	35, 6, 23, 40	35, 6, 22, 32	14, 29, 10, 39	10, 28,24	35, 2, 10, 31	10, 18, 39, 31	1, 29, 30, 36	3, 39, 18, 31	10, 13, 28, 38	14, 15, 18, 40	3, 36, 37, 10	29, 35, 3, 5	2, 14, 30, 40
24	Loss of information	10, 24, 35	10, 35, 5	1, 26	26	30, 26	30, 16		2, 22	26, 32				
25	Loss of time	10, 20, 37, 35	10, 20, 26, 5	15, 2, 29	30, 24, 14, 5	26, 4, 5, 16	10, 35, 17, 4	2, 5, 34, 10	35, 16, 32, 18		10, 37, 36,5	37, 36,4	4, 10, 34, 17	35, 3, 22, 5
26	Quantity of substance/the matter	35, 6, 18, 31	27, 26, 18, 35	29, 14, 35, 18		15, 14, 29	2, 18, 40, 4	15, 20, 29		35, 29, 34, 28	35, 14, 3	10, 36, 14, 3	35, 14	15, 2, 17, 40
27	Reliability	3, 8, 10, 40	3, 10, 8, 28	15, 9, 14, 4	15, 29, 28, 11	17, 10, 14, 16	32, 35, 40, 4	3, 10, 14, 24	2, 35, 24	21, 35, 11, 28	8, 28, 10, 3	10, 24, 35, 19	35, 1, 16, 11	
28	Measurement accuracy	32, 35, 26, 28	28, 35, 25, 26	28, 26, 5, 16	32, 28, 3, 16	26, 28, 32, 3	26, 28, 32, 3	32, 13, 6		28, 13, 32, 24	32, 2	6, 28, 32	6, 28, 32	32, 35, 13
29	Manufacturing precision	28, 32, 13, 18	28, 35, 27, 9	10, 28, 29, 37	2, 32, 10	28, 33, 29, 32	2, 29, 18, 36	32, 23, 2	25, 10, 35	10, 28, 32	28, 19, 34, 36	3, 35	32, 30, 40	30, 18
30	Object-affected harmful factors	22, 21, 27, 39	2, 22, 13, 24	17, 1, 39, 4	1, 18	22, 1, 33, 28	27, 2, 39, 35	22, 23, 37, 35	34, 39, 19, 27	21, 22, 35, 28	13, 35, 39, 18	22, 2, 37	22, 1, 3, 35	35, 24, 30, 18
31	Object-generated harmful factors	19, 22, 15, 39	35, 22, 1, 39	17, 15, 16, 22		17, 2, 18, 39	22, 1, 40	17, 2, 40	30, 18, 35, 4	35, 28, 3, 23	35, 28, 1, 40	2, 33, 27, 18	35, 1	35, 40, 27, 39
32	Ease of manufacture	28, 29, 15, 16	1, 27, 36, 13	1, 29, 13, 17	15, 17, 27	13, 1, 26, 12	16, 40	13, 29, 1, 40	35	35, 13, 8, 1	35, 12	35, 19, 1, 37	1, 28, 13, 27	11, 13, 1
33	Ease of operation	25, 2, 13, 15	6, 13, 1, 25	1, 17, 13, 12		1, 17, 13, 16	18, 16, 15, 39	1, 16, 35, 15	4, 18, 39, 31	18, 13, 34	28, 13 35	2, 32, 12	15, 34, 29, 28	32, 35, 30
34	Ease of repair	2, 27 35, 11	2, 27, 35, 11	1, 28, 10, 25	3, 18, 31	15, 13, 32	16, 25	25, 2, 35, 11	1	34, 9	1, 11, 10	13	1, 13, 2, 4	2, 35
35	Adaptability or versatility	1, 6, 15, 8	19, 15, 29, 16	35, 1, 29, 2	1, 35, 16	35, 30, 29, 7	15, 16	15, 35, 29		35, 10, 14	15, 17, 20	35, 16	15, 37, 1, 8	35, 30, 14
36	Device complexity	26, 30, 34, 36	2, 26, 35, 39	1, 19, 26, 24	26	14, 1, 13, 16	6, 36	34, 26, 6	1, 16	34, 10, 28	26, 16	19, 1, 35	29, 13, 28, 15	2, 22, 17, 19
37	Difficulty of detecting and measuring	27, 26, 28, 13	6, 13, 28, 1	16, 17, 26, 24	26	2, 13, 18, 17	2, 39, 30, 16	29, 1, 4, 16	2, 18, 26, 31	3, 4, 16, 35	30, 28, 40, 19	35, 36, 37, 32	27, 13, 1, 39	11, 22, 39, 30
38	Extent of automation	28, 26, 18, 35	28, 26, 35, 10	14, 13, 17, 28	23	17, 14, 13		35, 13, 16		28, 10	2, 35	13, 35	15, 32, 1, 13	18, 1
39	Productivity	35, 26, 24, 37	28, 27, 15, 3	18, 4, 28, 38	30, 7, 14, 26	10, 26, 34, 31	10, 35, 17, 7	2, 6, 34, 10	35, 37, 10, 2		28, 15, 10, 36	10, 37, 14	14, 10, 34, 40	35, 3, 22, 39

FIGURE 11.3
TRIZ contradiction matrix.

of brainstorming by providing a systematic approach that uses data and research.

As a tool for tactical innovation, TRIZ can be used in every phase of the define, measure, analyze, improve, and control (DMAIC) and DFSS methodologies to accelerate a team's creative problem solving. Within DFSS, it can be used in identify, design, optimize, validate (IDOV) or define, measure, analyze, design, and verify (DMADV). The following are several examples of where TRIZ can be used during DFSS:

	Improving feature / Worsening feature	Strength	Duration of action of moving object	Duration of action of stationary object	Temperature	Illumination intensity	Use of energy by moving object	Use of energy by stationary object	Power	Loss of energy	Loss of substance	Loss of information	Loss of time	Quantity of substance
		14	15	16	17	18	19	20	21	22	23	24	25	26
1	Weight of moving object	28, 27, 18, 40	5, 34, 31, 35		6, 29, 4, 38	19, 1, 32	35, 12, 34, 31		12, 36, 18, 31	6, 2, 34, 19	5, 35, 3, 31	10, 24, 35	10, 35, 20, 28	3, 26, 18, 31
2	Weight of stationary object	28, 2, 10, 27		2, 27, 19, 6	28, 19, 32, 22	19, 32, 35		18, 19, 28, 1	15, 19, 18, 22	18, 19, 28, 15	5, 8, 13, 30	10, 15, 35	10, 20, 35, 26	19, 6, 18, 26
3	Length of moving object	8, 35, 29, 34	19		10, 15, 19	32	8, 35, 24		1, 35	7, 2, 35, 39	4, 29, 23, 10	1,2	15, 2, 29	29, 35
4	Length of stationary object	15, 14, 28, 26		1, 10, 35	3, 35, 38, 18	3, 25			12, 8	6,28	10, 28, 24, 35	24, 26,	30, 29, 14	
5	Area of moving object	3, 15, 40, 14	6, 3		2, 15, 16	15, 32, 19, 13	19, 32		19, 10, 32, 18	15, 17, 30, 26	10, 35, 2, 39	30, 26	26, 4	29, 30, 6, 13
6	Area of stationary object	40		2, 10, 19, 30	35, 39, 38				17, 32	17, 7, 30	10, 14, 18, 39	30, 16	10, 35, 4, 18	2, 18, 40, 4
7	Volume of moving object	9, 14, 15, 7	6, 35, 4		34, 39, 10, 18	2, 13, 10	35		35, 6, 13, 18	7, 15, 13, 16	36, 39, 34, 10	2, 22	2, 6, 34, 10	29, 30, 7
8	Volume of stationary object	9, 14, 17, 15		35, 34, 38	35,6,4				30, 6		10, 39, 35, 34		35, 16, 32 18	35, 3
9	Speed	8, 3, 26, 14	3, 19, 35, 5		28, 30, 36, 2	10, 13, 19	8, 15, 35, 38		19, 35, 38, 2	14, 20, 19, 35	10, 13, 28, 38	13, 26		10, 19, 29, 38
10	Force (intensity)	35, 10, 14, 27	19, 2		35, 10, 21		19, 17, 10	1, 16, 36, 37	19, 35, 18, 37	14, 15	8, 35, 40, 5		10, 37, 36	14, 29, 18, 36
11	Stress or pressure	9, 18, 3, 40	19, 3, 27		35, 39, 19, 2		14, 24, 10, 37		10, 35, 14	2, 36, 25	10, 36, 3, 37		37, 36, 4	10, 14, 36
12	Shape	30, 14, 10, 40	14, 26, 9, 25		22, 14, 19, 32	13, 15, 32	2, 6, 34, 14		4, 6, 2	14	35, 29, 3, 5		14, 10, 34, 17	36, 22
13	Stability of the object's composition	17, 9, 15	13, 27, 10, 35	39, 3, 35, 23	35, 1, 32	32, 3, 27, 16	13, 19	27, 4, 29, 18	32, 35, 27, 31	14, 2, 39, 6	2, 14, 30, 40		35, 27	15, 32, 35
14	Strength	+	27, 3, 26		30, 10, 40	35, 19	19, 35, 10	35	10, 26, 35, 28	35	35, 28, 31, 40		29, 3, 28, 10	29, 10, 27
15	Duration of action of moving object	27, 3, 10	+		19, 35, 39	2, 19, 4, 35	28, 6, 35, 18		19, 10, 35, 38		28, 27, 3, 18	10	20, 10, 28, 18	3, 35, 10, 40
16	Duration of action by stationary object			+	19, 18, 36, 40				16		27, 16, 18, 38	10	28, 20, 10, 16	3, 35, 31
17	Temperature	10, 30, 22, 40	19, 13, 39	19, 18, 36, 40	+	32, 30, 21, 16	19, 15, 3, 17		2, 14, 17, 25	21, 17, 35, 38	21, 36, 29, 31		35, 28, 21, 18	3, 17, 30, 39
18	Illumination intensity	35, 19	2, 19, 6		32, 35, 19	+	32, 1, 19	32, 35, 1, 15	32	13, 16, 1, 6	13, 1	1, 6	19, 1, 26, 17	1, 19
19	Use of energy by moving object	5, 19, 9, 35	28, 35, 6, 18	-	19, 24, 3, 14	2, 15, 19	+	-	6, 19, 37, 18	12, 22, 15, 24	35, 24, 18, 5		35, 38, 19, 18	34, 23, 16, 18
20	Use of energy by stationary object	35				19, 2, 35, 32	-	+			28, 27, 18, 31			3, 35, 31
21	Power	26, 10, 28	19, 35, 10, 38	16	2, 14, 17, 25	16, 6, 19	16, 6, 19, 37		+	10, 35, 38	28, 27, 18, 38	10, 19	35, 20, 10, 6	4, 34, 19
22	Loss of energy	26			19,38,7	1, 13, 32, 15			3, 38	+	35, 27, 2, 37	19, 10	10, 18, 32, 7	7, 18, 25
23	Loss of substance	35, 28, 31, 40	28, 27, 3, 18	27, 16, 18, 38	21, 36, 39, 31	1, 6, 13	35, 18, 24, 5	28, 27, 12, 31	28, 27, 18, 38	35, 27, 2, 31	+		15, 18, 35, 10	6, 3, 10, 24
24	Loss of information		10	10		19			10,19	19,10		+	24, 26, 28, 32	24, 28, 35
25	Loss of time	29, 3, 28, 18	20, 10, 28, 18	28, 20, 10, 16	35, 29, 21, 18	1, 19, 26, 17	35, 38, 19, 18	1	35, 20, 10, 6	10, 5, 18, 32	35, 18, 10, 39	24, 26, 28, 32	+	35, 38, 18, 16
26	Quantity of substance/the matter	14, 35, 34, 10	3, 35, 10, 40	3, 35, 31	3, 17, 39		34, 29, 16, 18	3, 35, 31	35	7, 18, 25	6, 3, 10, 24	24, 28, 35	35, 38, 18, 16	+
27	Reliability	11, 28	2, 35, 3, 25	34, 27, 6, 40	3, 35, 10	11, 32, 13	21, 11, 27, 19	36, 23	21, 11, 26, 31	10, 11, 35	10, 35, 29, 39	10, 28	10, 30, 4	21, 28, 40, 3
28	Measurement accuracy	28, 6, 32	28, 6, 32	10, 26, 24	6, 19, 28, 24	6, 1, 32	3, 6, 32		3, 6, 32	26, 32, 27	10, 16, 31, 28		24, 34, 28, 32	2, 6, 32
29	Manufacturing precision	3, 27	3, 27, 40		19, 26	3, 32	32, 2		32, 2	13, 32, 2	35, 31, 10, 24		32, 26, 28, 18	32, 30
30	Object-affected harmful factors	18, 35, 37, 1	22, 15, 33, 28	17, 1, 40, 33	22, 33, 35, 2	1, 19, 32, 13	1, 24, 6, 27	10, 2, 22, 37	19, 22, 31, 2	21, 22, 35, 2	33, 22, 19, 40	22, 10, 2	35, 18, 34	35, 33, 29, 31
31	Object-generated harmful factors	15, 35, 22, 2	15, 22, 33, 31	21, 39, 16, 22	22, 35, 2, 24	19, 24, 39, 32	2, 35, 6	19, 22, 18	2, 35, 18	21, 35, 2, 22	10, 1, 34	10, 21, 29	1, 22	3, 24, 39, 1
32	Ease of manufacture	1, 3, 10, 32	27, 1, 4	35, 16	27, 26, 18	28, 24, 27, 1	28, 26, 27, 1	1, 4	27, 1, 12, 24	19, 35	15, 34, 33	32, 24, 18, 16	35, 28, 34, 4	35, 23, 1, 24
33	Ease of operation	32, 40, 3, 28	29, 3, 8, 25	1, 16, 25	26, 27, 13	13, 17, 1, 24	1, 13, 24		35, 34, 2, 10	2, 19, 13	28, 32, 2, 24	4, 10, 27, 22	4, 28, 10, 34	12, 35
34	Ease of repair	11, 1, 2, 9	11, 29, 28, 27	1	4, 10	15, 1, 13	15, 1, 28, 16		15, 10, 32, 2	15, 1, 32, 19	2, 35, 34, 27		32, 1, 10, 25	2, 28, 10, 25
35	Adaptability or versatility	35, 3, 32, 6	13, 1, 35	2, 16	27, 2, 3, 35	6, 22, 26, 1	19, 35, 29, 13		19, 1, 29	18, 15, 1	15, 10, 2, 13		35, 28	3, 35, 15
36	Device complexity	2, 13, 28	10, 4, 28, 15		2, 17, 13	24, 17, 13	27, 2, 29, 28		20, 19, 30, 34	10, 35, 13, 2	35, 10, 28, 29		6, 29	13, 3, 27, 10
37	Difficulty of detecting and measuring	27, 3, 15, 28	19, 29, 39, 25	25, 34, 6, 35	3, 27, 35, 16	2, 24, 26	35, 38	19, 35, 16	18, 1, 16, 10	35, 3, 15, 19	1, 18, 10, 24	35, 33, 27, 22	18, 28, 32, 9	3, 27, 29, 18
38	Extent of automation	25, 13	6, 9		26, 2, 19	8, 32, 19	2, 32, 13		28, 2, 27	23, 28	35, 10, 18, 5	35, 33	24, 28, 35, 30	35, 13
39	Productivity	29, 28, 10, 18	35, 10, 2, 18	20, 10, 16, 38	35, 21, 28, 10	26, 17, 19, 1	35, 10, 38, 19	1	35, 20, 10	28, 10, 29, 35	28, 10, 35, 23	13, 15, 23		35, 38

FIGURE 11.4
TRIZ contradiction matrix.

- Resolve inverse correlations in the roof (interaction matrix) during quality function deployment (see Figure 11.7).
- Develop alternate design concepts during conceptual design (see Figure 11.7).
- Develop new designs during the iterations of Pugh's concept selection matrix when conceptual designs are merged and improved.
- Resolve technical contradictions during design optimization.

	Improving feature ↓ / Worsening feature →	27 Reliability	28 Measurement accuracy	29 Manufacturing precision	30 Object-affected harmful factors	31 Object-generated harmful factors	32 Ease of manufacture	33 Ease of operation	34 Ease of repair	35 Adaptability or versatility	36 Device complexity	37 Difficulty of detecting and measuring	38 Extent of automation	39 Productivity
1	Weight of moving object	1, 3, 11, 27	28, 27, 35, 26	28, 35, 26, 18	22, 21, 18, 27	22, 35, 31, 39	27, 28, 1, 36	35, 3, 2, 24	2, 27, 28, 11	29, 5, 15, 8	26, 30, 36, 34	28, 29, 26, 32	26, 35 18, 19	35, 3, 24, 37
2	Weight of stationary object	10, 28, 8, 3	18, 26, 28	10, 1, 35, 17	2, 19, 22, 37	35, 22, 1, 39	28, 1, 9	6, 13, 1, 32	2, 27, 28, 11	19, 15, 29	1, 10, 26, 39	25, 28, 17, 15	2, 26, 35	1, 28, 15, 35
3	Length of moving object	10, 14, 29, 40	28, 32, 4	10, 28, 29, 37	1, 15, 17, 24	17, 15	1, 29, 17	15, 29, 35, 4	1, 28, 10	14, 15, 1, 16	1, 19, 26, 24	35, 1, 26, 24	17, 24, 26, 16	14, 4, 28, 29
4	Length of stationary object	15, 29, 28	32, 28, 3	2, 32, 10	1, 18		15, 17, 27	2, 25	3	1, 35	1, 26	26		30, 14, 7, 26
5	Area of moving object	29, 9	26, 28, 32, 3	2, 32	22, 33, 28, 1	17, 2, 18, 39	13, 1, 26, 24	15, 17, 13, 16	15, 13, 10, 1	15, 30	14, 1, 13	2, 36, 26, 18	14, 30, 28, 23	10, 26, 34, 2
6	Area of stationary object	32, 35, 40, 4	26, 28, 32, 3	2, 29, 18, 36	27, 2, 39, 35	22, 1, 40	40, 16	16, 4	16	15, 16	1, 18, 36	2, 35, 30, 18	23	10, 15, 17, 7
7	Volume of moving object	14, 1, 40, 11	25, 26, 28	25, 28, 2, 16	22, 21, 27, 35	17, 2, 40, 1	29, 1, 40	15, 13, 30, 12	10	15, 29	26, 1	29, 26, 4	35, 34, 16, 24	10, 6, 2, 34
8	Volume of stationary object	2, 35, 16		35, 10, 25	34, 39, 19, 27	30, 18, 35, 4	35		1		1, 31	2, 17, 26		35, 37, 10, 2
9	Speed	11, 35, 27, 28	28, 32, 1, 24	10, 28, 32, 25	1, 28, 35, 23	2, 24, 35, 21	35, 13, 8, 1	32, 28, 13, 12	34, 2, 28, 27	15, 10, 26	10, 28, 4, 34	3, 34, 27, 16	10, 18	
10	Force (Intensity)	3, 35, 13, 21	35, 10, 23, 24	28, 29, 37, 36	1, 35, 40, 18	13, 3, 36, 24	15, 37, 18, 1	1, 28, 3, 25	15, 1, 11	15, 17, 18, 20	26, 35, 10, 18	36, 37, 10, 19	2, 35	3, 28, 35, 37
11	Stress or pressure	10, 13, 19, 35	6, 28, 25	3, 35	22, 2, 37	2, 33, 27, 18	1, 35, 16	11	2	35	19, 1, 35	2, 36, 37	35, 24	10, 14, 35, 37
12	Shape	10, 40, 16	28, 32, 1	32, 30, 40	22, 1, 2, 35	35, 1	1, 32, 17, 28	32, 15, 26	2, 13, 1	1, 15, 29	16, 29, 1, 28	15, 13, 39	15, 1, 32	17, 26, 34, 10
13	Stability of the object's composition		13	18	35, 24, 30, 18	35, 40, 27, 39	35, 19	32, 35, 30	2, 35, 10, 16	35, 30, 34, 2	2, 35, 22, 26	35, 22, 39, 23	1, 8, 35	23, 35, 40, 3
14	Strength	11, 3	3, 27, 16	3, 27	18, 35, 37, 1	15, 35, 22, 2	11, 3, 10, 32	32, 40, 25, 2	27, 11, 3	15, 3, 32	2, 13, 25, 28	27, 3, 15, 40	15	29, 35, 10, 14
15	Duration of action of moving object	11, 2, 13	3	3, 27, 16, 40	22, 15, 33, 28	21, 39, 16, 22	27, 1, 4	12, 27	29, 10, 27	1, 35, 13	10, 4, 29, 15	19, 29, 39, 35	6, 10	35, 17, 14, 19
16	Duration of action by stationary object	34, 27, 6, 40	10, 26, 24		17, 1, 40, 33	22	35, 10	1	1	2		25, 34, 6, 35	1	20, 10, 16, 38
17	Temperature	19, 35, 3, 10	32, 19, 24	24	22, 33, 35, 2	22, 35, 2, 24	26, 27	26, 27	4, 10, 16	2, 18, 27	2, 17, 16	3, 27, 35, 31	26, 2, 19, 16	15, 28, 35
18	Illumination intensity		11, 15, 32	3, 32	15, 19	35, 19, 32, 39	19, 35, 28, 26	28, 26, 19	15, 17, 13, 16	15, 1, 19	6, 32, 13	32, 15	2, 26, 10	2, 25, 16
19	Use of energy by moving object	19, 21, 11, 27	3, 1, 32		1, 35, 6, 27	2, 35, 6	28, 26, 30	19, 35	1, 15, 17, 28	15, 17, 13, 16	2, 29, 27, 28	35, 38	32, 2	12, 28, 35
20	Use of energy by stationary object	10, 36, 23			10, 2, 22, 37	19, 22, 18	1, 4					19, 35, 16, 25		1, 6
21	Power	19, 24, 26, 31	32, 15, 2	32, 2	19, 22, 31, 2	2, 35, 18	26, 10, 34	26, 35, 10	35, 2, 10, 34	19, 17, 34	20, 19, 30, 34	19, 35, 16	28, 2, 17	28, 35, 34
22	Loss of energy	11, 10, 35	32		21, 22, 35, 2	21, 35, 2, 22		35, 32, 1	2, 19		7, 23	35, 3, 15, 23	2	28, 10, 29, 35
23	Loss of substance	10, 29, 39, 35	16, 34, 31, 28	35, 10, 24, 31	33, 22, 30, 40	10, 1, 34, 29	15, 34, 33	32, 28, 2, 24	2, 35, 34, 27	15, 10, 2	35, 10, 28, 24	35, 18, 10, 13	35, 10, 18	28, 35, 10, 23
24	Loss of information	10, 28, 23			22, 10, 1	10, 21, 22	32	27,22				35,33	35	13, 23, 15
25	Loss of time	10, 30, 4	24, 34, 28, 32	24, 26, 28, 18	35, 18, 34	35, 22, 18, 39	35, 28, 34, 4	4, 28, 10, 34	32, 1, 10	35, 28	6, 29	18, 28, 32, 10	24, 28, 35, 30	
26	Quantity of substance/the matter	18, 3, 28, 40	13, 2, 28	33, 30	35, 33, 29, 31	3, 35, 40, 39	29, 1, 35, 27	35, 29, 25, 10	2, 32, 10, 25	15, 3, 29	3, 13, 27, 10	3, 27, 29, 18	8, 35	13, 29, 3, 27
27	Reliability	+	32, 3, 11, 23	11, 32, 1	27, 35, 2, 40	35, 2, 40, 26		27, 17, 40	1, 11	13, 35, 8, 24	13, 35, 1	27, 40, 28	11, 13, 27	1, 35, 29, 38
28	Measurement accuracy	5, 11, 1, 23	+		28, 24, 22, 26	3, 33, 39, 10	6, 35, 25, 18	1, 13, 17, 34	1, 32, 13, 11	13, 35, 2	27, 35, 10, 34	26, 24, 32, 28	28, 2, 10, 34	10, 34, 28, 32
29	Manufacturing precision	11, 32, 1		+	26, 28, 10, 36	4, 17, 34, 26		1, 32, 35, 23	25, 10		26, 2, 18		26, 28, 18, 23	10, 18, 32, 39
30	Object-affected harmful factors	27, 24, 2, 40	28, 33, 23, 26	26, 28, 10, 18	+		24, 35, 2	2, 25, 28, 39	35, 10, 2	35, 11, 22, 31	22, 19, 29, 40	22, 19, 29, 40	33, 3, 34	22, 35, 13, 24
31	Object-generated harmful factors	24, 2, 40, 39	3, 33, 26	4, 17, 34, 26		+					19,1,31	2, 21, 27, 1	2	22, 35, 18, 39
32	Ease of manufacture		1, 35, 12, 18		24, 2		+	2, 5, 13, 16	35, 1, 11, 9	2, 13, 15	27, 26, 1	6, 28, 11, 1	8, 28, 1	35, 1, 10, 28
33	Ease of operation	17, 27, 8, 40	25, 13, 2, 34	1, 32, 35, 23	2, 25, 28, 39		2, 5, 12	+	12, 26, 1, 32	15, 34, 1, 16	32, 26, 12, 17		1, 34, 12, 3	15, 1, 28
34	Ease of repair	11, 10, 1, 16	10, 2, 13	25, 10	35, 10, 2, 16		1, 35, 11, 10	1, 12, 26, 15	+	7, 1, 4, 16	35, 1, 13, 11		34, 35, 7, 13	1, 32, 10
35	Adaptability or versatility	35, 13, 8, 24	35, 5, 1, 10		35, 11, 32, 31		1, 13, 31	15, 34, 1, 16	1, 16, 7, 4	+	15, 29, 37, 28	1	27, 34, 35	35, 28, 6, 37
36	Device complexity	13, 35, 1	2, 26, 10, 34	26, 24, 32	22, 19, 29, 40	19, 1	27, 26, 1, 13	27, 9, 26, 24	1, 13	29, 15, 28, 37	+	15, 10, 37, 28	15, 1, 24	12, 17, 28
37	Difficulty of detecting and measuring	27, 40, 28, 8	26, 24, 32, 28		22, 19, 29, 28	2, 21	5, 28, 11, 29	2, 5	12, 26	1, 15	15, 10, 37, 28	+	34, 21	35, 18
38	Extent of automation	11, 27, 32	28, 26, 10, 34	28, 26, 18, 23	2, 33	2	1, 26, 13	1, 12, 34, 3	1, 35, 13	27, 4, 1, 35	15, 24, 10	34, 27, 25	+	5, 12, 35, 26
39	Productivity	1, 35, 10, 38	1, 10, 34, 28	18, 10, 32, 1	22, 35, 13, 24	35, 22, 18, 39	35, 28, 2, 24	1, 28, 7, 10	1, 32, 10, 25	1, 35, 28, 37	12, 17, 28, 24	35, 18, 27, 2	5, 12, 35, 26	+

FIGURE 11.5
TRIZ contradiction matrix.

- Resolve contradictions between input parameters in the transfer function.
- Diagnose failures (perform failure analysis) that could not be resolved using failure mode and effects analysis (FMEA) through reverse TRIZ.

Inventive principles
1. Segmentation
2. Extraction, separation, removal, segregation
3. Local quality
4. Asymmetry
5. Combining, integration, merging
6. Universality, multifunctionality
7. Nesting
8. Counterweight, levitation
9. Preliminary anti-action, prior counteraction
10. Prior action
11. Cushion in advance, compensate before
12. Equipotentiality, remove stress
13. Inversion, the other way around
14. Spheroidality, curvilinearity
15. Dynamicity, optimization
16. Partial or excessive action
17. Moving to a new dimension
18. Mechanical vibration/oscillation
19. Periodic action
20. Continuity of a useful action
21. Rushing through
22. Convert harm into benefit, "blessing in disguise"
23. Feedback
24. Mediator, intermediary
25. Self-service, self-organization
26. Copying
27. Cheap, disposable objects
28. Replacement of a mechanical system with "fields"
29. Pneumatics or hydraulics:
30. Flexible membranes or thin film
31. Use of porous materials
32. Changing color or optical properties
33. Homogeneity
34. Rejection and regeneration, discarding and recovering
35. Transformation of the physical and chemical states of an object, parameter change, changing properties
36. Phase transformation
37. Thermal expansion
38. Use strong oxidizers, enriched atmospheres, accelerated oxidation
39. Inert environment or atmosphere
40. Composite materials

FIGURE 11.6
The 40 principles of invention.

Conclusion

Six Sigma and DFSS were developed to generate breakthrough improvement. To achieve breakthrough improvement, innovation is necessary. For organizations to achieve aggressive goals such as a 95% reduction in defects or a 20% increase in market share, a structured innovation process with the involvement of all employees is critical. Likewise, for innovation to be successful, the Six Sigma philosophies of variation reduction and continuous improvement must be in place. Innovation and Six Sigma are compatible and should be used in parallel. For this to happen, leadership must provide the appropriate conditions for the methods to be used synergistically. In the next chapter, Lean design is discussed to reduce or eliminate waste in product development.

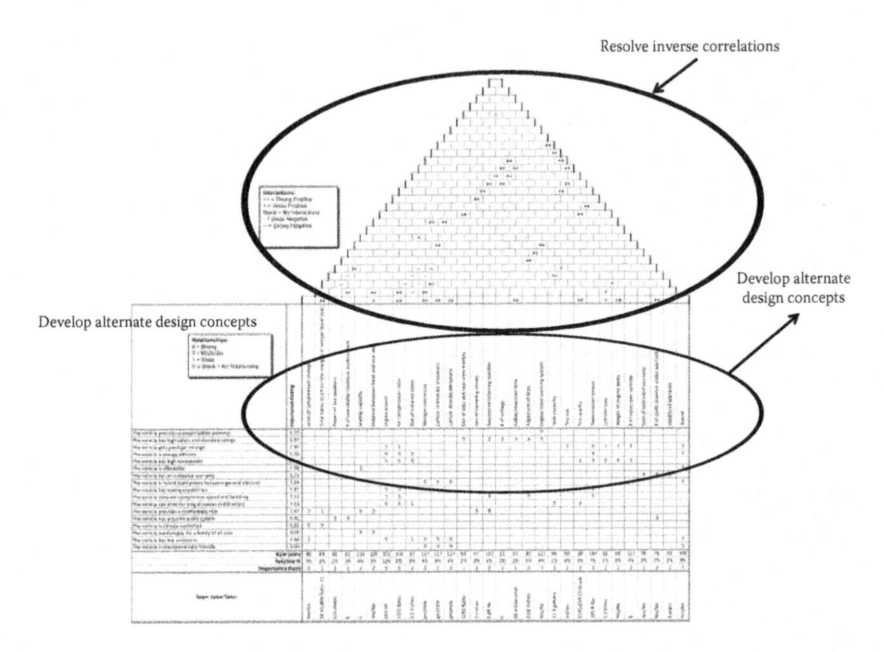

FIGURE 11.7
TRIZ in quality function deployment.

Questions

1. What is TRIZ?
2. What is the general concept behind TRIZ?
3. What is the degree of ideality?
4. What are the benefits of TRIZ?
5. What does it mean when a system is robust?
6. What is Nine Windows?
7. What is a technical contradiction? Give an example.
8. What is a physical contradiction? Give an example.
9. What is the separation in time principle? Give an example.
10. When should the contradiction matrix be used?
11. When should TRIZ be used in DFSS?

12

Lean Design

Lean is a compilation of several diverse methods and tools that provide a more holistic and integrated toolkit for process improvement. These same tools can be applied for designing and developing products and services. As new products and services become increasingly complex and multifaceted, it is necessary to more tightly couple and integrate the entire service and product development process with Lean. A comprehensive methodology is necessary utilizing Lean techniques to design a product (application), process, or service right the first time.

The Lean Six Sigma family includes Lean, Six Sigma, and Design for Six Sigma (DFSS). The successful integration of each component will yield improvements. The components can be seen in Figure 12.1.

Lean consists of eliminating wastes to create better processes, whereas Six Sigma focuses on reducing defects and minimizing variation. Lean is primarily used when there is waste, inventory, and redundancy in processes. Lean also focuses on workflow and the elimination of human error while making processes more efficient.

Lean concepts also consist of continuous flow and pull methodologies. Lean focuses on cost, delivery, quality, safety, and most importantly the people. The concept of Lean is to ensure customer satisfaction, reduce internal and external wastes, increase capacity, simplify operations, reduce rework, and remain competitive.

DFSS is used when design issues are apparent. When an increased amount of process variation occurs in complex processes, DFSS is the proper tool to use. DFSS incorporates the use of root cause analysis for challenging processes. DFSS is also used when there is a high degree of technical complexity.

Lean DFSS empowers employees to create process stability while promoting a cultural awareness of continuous improvement. The elements include systematic problem-solving techniques. Through data collection and analysis of the data, a reliable process can be captured for Lean DFSS. Lean DFSS should include the following elements:

- Specify value by product
- Identify the value stream
- Make value flow
- Let the customer pull the value from the producer
- The pursuit of perfection

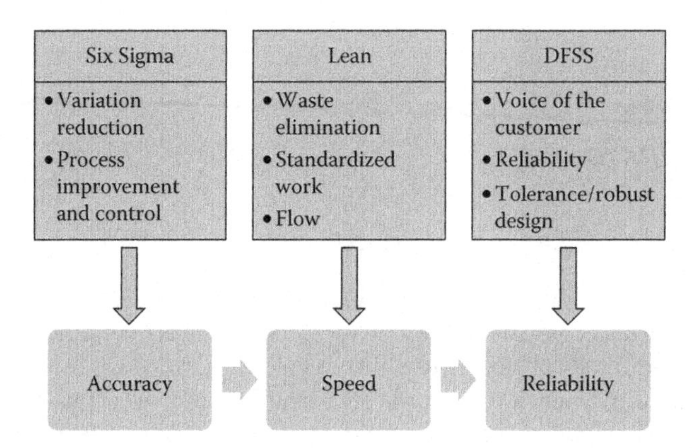

FIGURE 12.1
Lean Six Sigma and DFSS.

A concept called *value stream mapping* is performed to identify all of the activities involved in the processes from beginning to end including suppliers and end customers. Value stream mapping is conducted for a visual representation of what is occurring while being able to capture the "big picture." The focus should be on material and informational flows and identifying improvement opportunities.

The steps for value stream mapping are as follows:

- Map current state processes
- Summarize current state processes
- Map future state processes
- Summarize future state processes
- Develop short-term plans to move from the current to future state
- Develop long-term plans to move from the current to future state
- Implement risks, failures, and processes for transitions
- Map key project owners, key dates, and future dates of future states
- Continue to map new future states when future states are met

The main symbols for value stream mapping are shown in Figure 12.2.

Value stream mapping is a key component to establish that all steps occurring are adding value. Value-added activities are any activity that the customer is willing to pay for. Another note to remember is to not only have a smart and efficient technique, but also only produce goods that the customer is demanding to eliminate excess inventory.

This technique is called the *pull technique*. Pull is the practice of not producing any goods upstream if the downstream customer does not need it. The reason this is a difficult technique is because once an efficient method

Symbols	Description	Symbols	Description	Symbols	Description
	Customer/ Supplier Start or end point for material flow	Kaizen	Kaizen blitz Area for improvement		External shipment Shipments to or from suppliers
Process box Shared process box	Process Machine operation or department through which material flows		Supermarket Small inventory for immediate production	FIFO	FIFO First in, first out lane
P/T = 15 L/T = 15 %C&A = 99% Rel = 95%	Data box		Buffer Safety stock		In box Information queues
100 units 100 units	Inventory	Pull	Pull symbols Replenish stock in supermarket		Internal movement
Work cell	Work cell	Kanban	Kanban card Replenish stock in supermarket		People, phones, operators, etc.
	Push arrow	OXOX	Load leveling		
1 day	NVA delay		Go and see When there is a problem, go and see what's wrong.	MRP/ Scheduler	Scheduling

FIGURE 12.2
Value stream mapping symbols.

is found to produce goods, mass production begins. The operations forget if the goods are actually needed or not and begin thinking only of throughput. Even though co-manufacturers seem like a bad idea for many employers, they are sometimes come in handy when a small amount of a versatile product is needed.

Push systems on the other hand are not effective due to predictions of customer demands.

Lean systems show the pull system utilizing machinery for 90% of requirements and limits downtime to 10% for changeovers and maintenance. This does not mean preventive maintenance should not be performed, but only that the maintenance time is reduced to 10%. Kanbans are a key factor in a Lean system since they provide a visual indicator that another part or process is required. This also prevents excess parts from being made or excess processes being performed.

Heijunka is the leveling of production and scheduling based on volume and product mix. Instead of building products according to the flow of customer orders, this technique levels the total volume of orders over a specific time so that uniform batches of different product mixes are made daily. The result is a predictable matrix of product types and volumes For heijunka to succeed, changeovers must be managed easily. Changeovers must be as minimally invasive as possible to prevent time wasted because of product mix. Another key to heijunka is making sure that products are needed by customers. A product should not be included in a mix simply to produce inventory if it is not demanded by customers. Long changeovers should be investigated to determine the reason and devise a method to shorten them.

Single-Minute Exchange of Dies (SMED)

What Is SMED?

SMED consists of the following:

- Theory and set of techniques to make it possible to perform equipment setup and changeover operations in under 10 minutes
- Originally developed to improve die press and machine tool setups, but principles apply to changeovers in all processes
- It may not be possible to reach the "single-minute" range for all setups, but SMED dramatically reduces setup times in almost *every* case
- Leads to benefits for the company by giving customers a variety of products in just the quantities they need
- High quality, good price, speedy delivery, less waste, cost-effective

It is important to understand large-lot production, which can lead to issues.

The three key topics to consider when understanding large-lot production are the following:

- Inventory waste
  - Storing what is not sold costs money
  - Ties up company resources with no value to the product
- Delay
  - Customers have to wait for the company to produce entire lots rather than just what they want

- Declining quality
  - Storing unsold inventory increases chances of product being scrapped or reworked, which adds costs

Once this is realized, the benefits of SMED can be understood:

- Flexibility
  - Meet changing customer needs without excess inventory
- Quicker delivery
  - Small-lot production equals less lead time and less customer waiting time
- Better quality
  - Less inventory storage equals fewer storage-related defects
  - Reduction of setup errors and elimination of trial runs for new products
- Higher productivity
  - Reduction in downtime
  - Higher equipment productivity rate

Two types of operations are realized during setup operations, which consist of internal and external operations. Internal setup is a setup that can only be performed when the machine is shut down (i.e., a new die can only be attached to a press when the press is stopped).

External setup is a setup that can be performed while the machine is still running (i.e., bolts attached to a die can be assembled and sorted while the press is operating).

It is important to convert as much internal work as possible to external work, which is shown in Figure 12.3. A pull-flow diagram is shown in Figure 12.4.

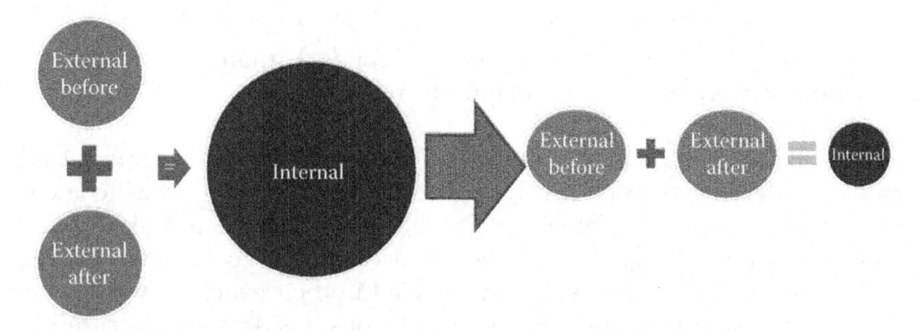

FIGURE 12.3
SMED internal vs. external.

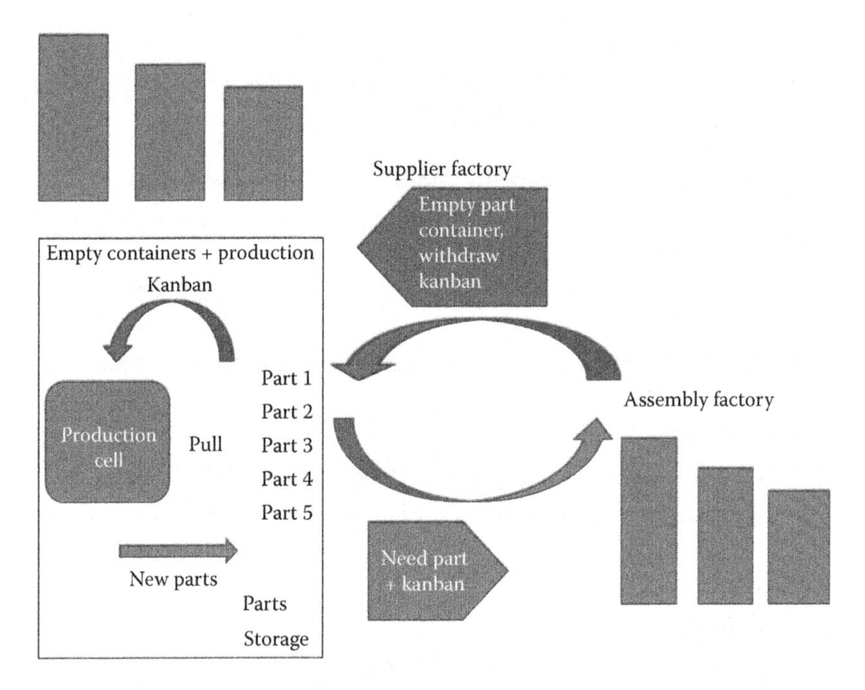

FIGURE 12.4
Pull-flow design.

The Lean action plan is simply drawn out in five steps:

1. Getting started: Plan out the appropriate steps. This will take 1–6 months.
2. Create the new organization and restructure. This will take 6–24 months.
3. Implement Lean techniques and systems and continually improve. This will take two to four years.
4. Complete the transformation. This will take up to five years.
5. Perform the entire process again to conduct another continuous improvement project and sustain the results.

What also cannot be lost through this process is top management leadership. Leaders need to be involved through each step in order to incorporate their ideas with employees' feedback. Leaders should also have a vision of what the end goal is and should cascade their goals down. Leaders help employees to be successful and satisfied in their everyday work while empowering them to make influential decisions. Leaders should carefully plan and implement strategically while assigning the proper resources. The focus should be on the right priorities, which is also the concept of the vital

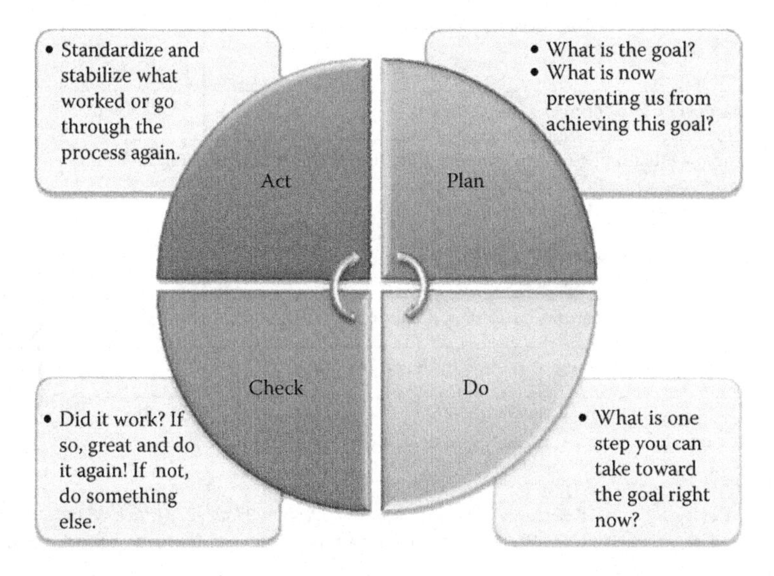

FIGURE 12.5
Plan do check act.

few over the trivial many. The projects should be based on organizational goals for the business. The projects should be selected based on risk analysis. The focus should cascade down to the customers' needs, which include the operations, the business, and the processes. Consistency through the organization will help the strategies be successful.

The plan, do, check, act methodology should be used during this implementation, which can be seen in Figure 12.5.

The three aspects for improving the product development process include:

1. Maximize profitability
2. Minimize time
3. Minimize cost

This must be balanced without compromising value from the customer's perspective.

Lean design/product development aids in identifying and reducing or eliminating waste in the product development process. Lean design focuses on removing waste from all aspects of the product and the associated development process before the start of manufacture. Lean design addresses the entire life cycle of a product. More specifically, Lean design targets cutting manufacturing costs during the design cycle and accelerating the time-to-market. Kearney (2003) identified the most common forms of waste in product design as shown in Figure 12.6.

Area of waste reduction	% of design waste	
Designs never used, completed, or delivered	Unknown	
Downtime while finding information, waiting for test results, etc.	33%–50%	
Unnecessary documents and prototypes		
Underutilization of design knowledge, for example in costly parts	18%	58%
Over design, such as features customers do not need	8%	
Validating manufacturing errors early in the design process	17%	
Poor designs producing product defects	15%	

FIGURE 12.6
Waste in design. (From Kearney, A.T., *The Line on Design: How to Reduce Material Cost by Eliminating Design Waste,* UGS PLM Solutions analysis of Tier 1 Automotive suppliers, http://www.atkearney.com, 2003.)

Mascitelli (2004) developed five principles of Lean design:

Principle 1: Precisely define the customer's *problem* and identify the specific *functions* that must be performed to solve that problem.

Principle 2: Identify the *fastest process* by which the identified functions can be integrated into a high-quality, low-cost product.

Principle 3: Strip away any *unnecessary* or *redundant* cost items to reveal the optimal product solution.

Principle 4: Listen to the voice of the customer *frequently* and *iteratively* throughout the development process.

Principle 5: Embed cost-reduction tools and methods into both your *business practices* and your *culture* to enable cost reduction.

The most basic concept when discussing waste reduction begins with kaizen. Kaizen is a Japanese concept defined as "taking apart and making better." The concept takes a vast amount of project-management techniques to facilitate the process going forward. 5s processes are the most predominant and commonly known for kaizen events.

5s principles are determined by finding a place for everything and everything in its place The 5s levels are as follows:

Sort: Identify and eliminate items, and dispose of unneeded materials that do not belong in an area. This reduces waste, creates a safer work area, opens space, and helps visualize processes. It is important to sort through the entire area. The removal of items should be discussed with all personnel involved. Items that cannot be removed immediately should be tagged for subsequent removal.

Sweep: Clean the area so that it looks like new and clean it continuously. Sweeping prevents an area from getting dirty in the first place and eliminates further cleaning. A clean workplace indicates high standards of quality and good process controls. Sweeping should eliminate dirt, build pride in work areas, and build value in equipment.

Straighten: Have a place for everything and everything in its place. Arranging all necessary items is the first step. It shows what items are required and what items are not in place. Straightening aids efficiency; items can be found more quickly and employees travel shorter distances. Items that are used together should be kept together. Labels, floor markings, signs, tape, and shadowed outlines can be used to identify materials. Shared items can be kept at a central location to eliminate purchasing more than needed.

Schedule: Assign responsibilities and due dates to actions. Scheduling guides sorting, sweeping, and straightening, and prevents regressing to unclean or disorganized conditions. Items are returned where they belong and routine cleaning eliminates the need for special cleaning projects. Scheduling requires checklists and schedules to maintain and improve neatness.

Sustain: Establish ways to ensure maintenance of manufacturing or process improvements. Sustaining maintains discipline. Utilizing proper processes will eventually become routine. Training is key to sustaining the effort and involvement of all parties. Management must mandate the commitment to housekeeping for this process to be successful.

The benefits of 5s include (a) a cleaner and safer workplace; (b) customer satisfaction through better organization; and (c) increased quality, productivity, and effectiveness.

Kai is defined as to break apart or disassemble so that one can begin to understand. Zen is defined as to improve. This process focuses on improvements objectively by breaking down the processes in a clearly defined and understood manner so that wastes are identified, improvement ideas are created, and wastes are both identified and eliminated. The philosophy includes reducing cycle times and lead times, in turn increasing productivity, reducing work in process (WIP), reducing defects, increasing capacity, increasing flexibility, and improving layouts through visual management techniques.

Operator cycle times need to be understood in order to reduce the non-productive times. Operators should also be cross-functional so that they are able to perform different job functions and the workloads of each function are well balanced. The work performed needs to be not only value-added work, but also meet customer needs and expectations. WIP should be eliminated to reduce inventory. Inventory should be seen simply as money waiting in process and should be reduced as much as possible. WIP can be reduced by reducing setup times, transporting smaller quantities of batch outputs, and line balancing. Bottlenecks should be removed by finding non-value-added tasks and removing the excess time spent by both machinery and people.

Key areas of waste in product design stem from process delays, design reuse, defects, and process efficiency. Process delays are caused by time lost looking for information, waiting for test results, and waiting for feedback. Waste from design reuse is due to not learning from past design experiences, not reducing unnecessary features, and designs that are never used, completed, or delivered. Defect wastes stem from poor designs and warranty issues. Finally, process efficiency waste is caused by underutilization of design knowledge and not validating manufacturing errors early.

Mistake proofing is a subject of its own when brought into the Lean Six Sigma methodology. This term, often called *poka yoke,* is also another initiative to improve production systems. The methodology eliminates product defects before they occur by installing processes to prevent the mistakes from happening in the first place. The mistakes that happen are due to human nature, and can normally not be eliminated by simple training or standard operating procedures (SOPs). The steps to eliminate the defect will prevent the next step in the process from occurring if a defect is found. Normally, there is some type of alert that will show there is a mistake and will prevent the process from going forward. An example of a poka yoke would be a simple check weigher that would kick off a package of food if it is not the correct weight.

Poka yokes often also encompass a concept called zero quality control (ZQC). This does not mean a reduction in defects, but instead the complete elimination of defects, also known as *zero defects.* ZQC is another Japanese concept that leads to low-inventory production. The inventory is low because defective parts are made less often. ZQC also focuses on quality control and data versus blaming people for mistakes. The methodology was developed by Shigeo Shingo who understood it was human nature to make common mistakes and did not feel people should be reprimanded for them. Shingo said, "Punishment makes people feel bad, it does not eliminate defects."

Takt time is a kaizen tool used in the order-taking phase. *Takt* is a German word for pace. Takt time is defined as time per unit. This is the operational measurement to keep production on track. To calculate Takt time, the formula is time available/production required. Thus, if a required production is 100 units per day and 240 minutes are available, the Takt time = 240/100 or 2.4 minutes to keep the process on track. Individual cycle times should be balanced to the pace of Takt time. To determine the number of employees required, the formula is (labor time/unit)/Takt time. Takt in this case is time per unit. Takt requires visual controls and helps reduce accidents and injuries in the workplace. Monitoring inventory and production WIP will reduce waste or muda. *Muda* is a Japanese term for waste where waste is defined as any activity that consumes some type of resource but is non-value-added for the customer. The customer is not willing to pay for this resource because it is not benefiting them. Types of muda include scrap, rework, defects, mistakes, and excess transport, handling, or movement.

This concept is important because it focuses on the customers and realizes that defects are costly; therefore, eliminating defects saves money. Many companies "rework" products to save money, but do not think to eliminate the problem in the first place. This process will eliminate rework by eliminating any defects from happening in the first place.

There are several ways to decrease costs in the design cycle by reducing direct material cost, direct labor cost, operational overhead, nonrecurring design cost, and product-specific capital investments. Direct material costs can be reduced by using common parts, design simplification, defect reduction, and parts-count reduction. Direct labor costs can be reduced through design simplification, design for manufacture and assembly, and standardizing processes. Operational overheads can be reduced by increasing the utilization of shared capital equipment and modular design. Nonrecurring design costs can be decreased by standardization, value engineering, and platform design strategies. Product-specific capital investments can be minimized by using value engineering, part standardization, and one-piece flow.

Huthwaite (2004) developed five laws of Lean design, which include:

1. Law of strategic value: Ensure you are delivering value to all stakeholders during the product's life cycle.
2. Law of waste prevention: Prevent waste in all aspects of the product's life.
3. Law of marketplace pull: Anticipate change in order to deliver the right products at the right time.
4. Law of innovation flow: Create new ideas to delight customers and differentiate your product.
5. Law of last feedback: Use predictive feedback to forecast cause and effect relationships.

By incorporating Lean principles into product and process design, further improvements can be made to the design of a product or service. The product development process can be shortened, bringing the product to market faster while still ensuring value to the customer.

Conclusion

Lean is a process-improvement methodology that can be applied for designing and developing products and services to reduce and/or eliminate waste. As products become more complex, Lean techniques can be utilized to design a product (application), process, or service right the first time. Design for X is covered in Chapter 13 and presents methodologies for designing for certain attributes such as reliability, manufacturability, and the environment.

Questions

1. What is Lean?
2. What are the elements of Lean DFSS?
3. How can Lean be applied to product development?
4. What is the difference between pull and push systems? Give an example.
5. What is heijunka? Give an example.
6. What are the most common forms of waste in product design?
7. Give an example of defect waste in product design.
8. How can nonrecurring design costs be reduced?

References

Huthwaite, B. (2004). *The Lean Design Solution*. Institute for Lean Design. Mackinac Island, MI.

Kearney, A.T. (2003). *The Line on Design: How to Reduce Material Cost by Eliminating Design Waste*, UGS PLM Solutions analysis of Tier 1 Automotive suppliers, http://www.atkearney.com.

Mascitelli, R. (2004). *The Lean Design Guidebook*. Technology Perspectives. Northridge, CA.

13

Design for X Methods

Decoration is just make-up for the wrinkles of the idea.

Thomas Mann

Design for X (DFX) involves designing in attributes such as

- Manufacturability (also known as produceability)
- Assembly
- Serviceability (also known as maintenance)
- Reliability
- Testability
- Environment

DFX is also often referred to as the "design for the ___ilities."

Design for Manufacturability

Design for manufacturability (DFM) is a philosophy that strives to improve costs and employee safety by simplifying the manufacturing process through product design. The purpose of DFM is to design products during the conceptual stage so that they are easy to manufacture. This reduces or eliminates redesign. It is specifically concerned with materials and manufacturing processes. As part of DFM, it is important to involve the production and supply chain.

Design for Assembleability

Design for assembleability (DFA) is closely linked to DFM. However, it is concerned with the ability to assemble a product by hand or through automation. The purpose is to simplify the assembly steps. The main assembleability aspects include

- Ease of assembly motions (should be ergonomical)
- Number of assembly motions
- Number of components (reduce the number of similar components)
- Complexity of component interfaces
- Number of assembly interfaces
- Type of fastening technology
- Number of fasteners or fastening functions

Design for Reliability

Design for reliability is the probability that a system will perform its intended function under the conditions stated without failing for a specific amount of iterations. Failure is the ability for the system to malfunction or not meet the proper functionality for which it is intended. The quality component consists of the entirety of the features and characteristics of the product or service to ensure it meets its intended use. In short, reliability consists of how well a product or service performs over time. In order for a process to meet reliability, the following must be in place:

- Proper specifications
- Critical-to-quality parameters
- Advanced quality planning
- Design functionality
- Supplier quality
- Manufacturing quality
- Serviceability
- Feedback for improvements

Reliability is covered in more detail in Chapter 19.

Design for Serviceability

The purpose of design for serviceability (DFS) is to ensure that product concepts are easy to service and maintain. DFS is linked to DFA because the easier it is to assemble a product, the easier it is to disassemble a product for service and maintenance. Therefore, it is critical to involve maintenance and service personnel in the design. The key focus areas for DFS include

- Accessibility
- Low fastener count
- Low tool count
- One-step service functions
- Predictable maintenance schedules

Design for Environment

The key purpose of design for the environment (DFE) is to develop reusable or recyclable elements. When designing new products and processes, the team should look for environmentally friendly solutions.

The limitations of DFX are when environmental causes are not included in simulations; therefore, best-case scenarios are predicted.

Noise factors can originate with your suppliers, in manufacturing, or in your customer's usage environment.

Design for Testability

When designing products that will require advanced testing, the teams should consider the measurement system and data acquisition system during the conceptual design. The measurement system should measure the ideal/transfer function for the component or system. This should be clearly defined for each concept. The measurement should be a continuous variable.

Conclusion

DFX enables the development of new products, processes, or services based on the attributes that are critical to customer satisfaction. The DFX principles should be applied to each level of the system architecture including the system, subsystem, subassembly, component, and manufacturing process. Once the team has determined the DFX characteristics, the design team can then develop multiple designs to meet the customers' needs. Chapter 14 discusses concept generation to develop these various designs and then a methodology for selecting the best concept based on the customer criteria.

Questions

1. What is DFX?
2. What is the purpose of design for manufacturability?
3. What are the main aspects of assembleability?
4. What is design for reliability?
5. How is design for serviceability related to design for assembleability?
6. What happens when environmental causes are not included in simulations?

14

Pugh Concept Selection Matrix

Becoming limitless involves mental agility; the ability to quickly grasp and incorporate new ideas and concepts with confidence.

Lorli Myers

Concept generation can come from several sources. First, concepts can stem from personal inspiration and vision. Concept ideas can also be a result of benchmarking the industry leaders or reverse engineering the best-in-class products based on their form, fit, or function. Another methodology for concept generation is TRIZ, which is discussed in detail in Chapter 11. Brainstorming with a cross-functional group can also be beneficial for developing new ideas and concepts. Finally, concepts can be generated through patent searches, literature searches, and consultant input and guidance.

The purpose of a Pugh Concept Selection Matrix (PCSM) is to evaluate several alternatives in order to select the best alternative against a certain set of specific criteria. It is a structured concept selection process used by multidisciplinary teams to converge on superior concepts. The methodology uses a matrix consisting of criteria based on the voice of the customer (VOC) and its relationship to specific, candidate design concepts. Then evaluations are made by comparing new concepts with a benchmark called the *datum*.

The uses of a PCSM are to evaluate design concepts for a new product or process against a prioritized list of critical to quality (CTQ) characteristics. Important decisions such as make versus buy, supply chain operations, purchasing of expenditures for capital investments, and so forth can be made within a PCSM by evaluating alternatives. The PCSM must be performed as a team and should not be conducted alone since the results could be negatively skewed. The concept of the PCSM consists of four main steps as shown in Figure 14.1. Figures 14.2 and 14.3 provide an example of a PCSM. The weighted PCSM is the next step, an example of which is shown in Figure 14.4.

The design trade-offs need to be discussed next after the weighted PCSM. The purpose of the design trade-offs is to evaluate several alternatives quantitatively in order to select the best alternative against a specific set of criteria. This process is similar to a PCSM, except it is less qualitative. The design trade-off matrix is a powerful tool because it allows

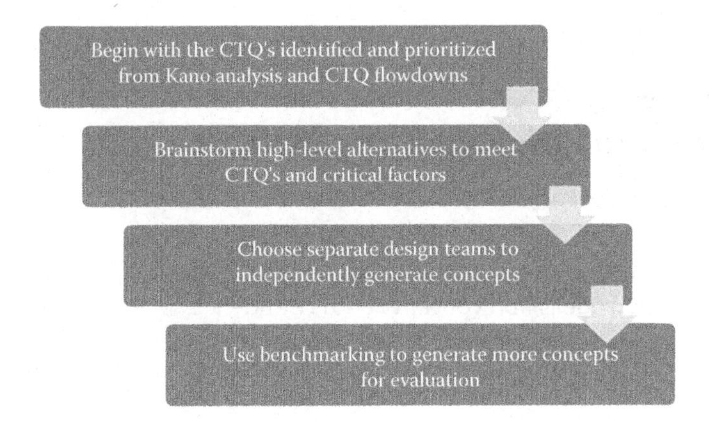

FIGURE 14.1
Pugh Concept Selection Matrix steps.

Criteria	Rating	Datum	Concepts 1	2	3	4	5	6	7
Safe operation	5		S	S	S	S	S	S	S
Self-cleaning teeth	2		+	+	–	+	S	+	–
Containment of shredded particles	3		+	S	S	+	+	S	+
Easy to empty	3		+	S	–	+	+	S	S
Overfill prevention	4		S	S	S	S	S	S	S
Compaction of particles	1		S	S	S	S	S	S	+
Single-hand usage	2		S	S	S	–	S	S	S
Less jamming	4		–	S	–	+	–	S	–
Sum of (+)			3	1	0	4	2	1	2
Sum of (–)			1	0	3	1	1	0	2
Sum of (S)			4	7	5	3	5	7	4
Weighted sum of positives (+)			8	2	0	12	6	2	4
Weighted sum of negatives (–)			4	0	9	2	4	0	6

FIGURE 14.2
Pugh Matrix.

the best selection of each of the concepts when there are several models or simulations that can be used to calculate or predict the actual values for the CTQs. It is also an excellent way to justify financial investments where several financial analyses can be performed and compared against alternatives.

The design trade-off matrix is shown in Figure 14.5.

The PCSM is a structured approach that enables a team to converge on a superior concept. The matrix includes the criteria from the VOC to evaluate candidate design concepts. The methodology is used to evaluate new

Pugh Concept Selection Matrix

	1 to 5	Concepts Note: Enter "+" or "-" Select a Baseline Concept as Reference Datum																			
Criteria	Rating	1	2	3	4	5	6	7	8	9	10	11	12	13	14	15	16	17	18	19	20
Speed	4	-	s	s																	
Sensor detection	4	+	s	s																	
Small # of subassy.	3	+	s	s																	
Time of installation	3	s	+	s																	
Dimension of subassy	3	s	s	s																	
MTBF	5	+	s	s																	
Movement accuracy	5	s	-																		
Implementation time	5	+	-	-																	
CTQ 9																					
CTQ 10																					
CTQ 11																					
CTQ 12																					
CTQ 13																					
CTQ 14																					
CTQ 15																					
CTQ 16																					
CTQ 17																					
CTQ 18																					
note: insert rows above this line to maintain calculations																					
Sum of Positives (+)		4	1	0	0	0	0	0	0	0	0	0	0	0	0	0	0	0	0	0	0
Sum of Negatives (-)		1	2	1	0	0	0	0	0	0	0	0	0	0	0	0	0	0	0	0	0
Sum of Sames (S)		3	5	6	0	0	0	0	0	0	0	0	0	0	0	0	0	0	0	0	0
Weighted Sum of Positives		17	3	0	0	0	0	0	0	0	0	0	0	0	0	0	0	0	0	0	0
Weighted Sum of Negatives		-4	-10	-5	0	0	0	0	0	0	0	0	0	0	0	0	0	0	0	0	0

FIGURE 14.3
Pugh Concept Selection Matrix.

Weighted Pugh Selection Matrix

Concepts

Enter "-3" to "+3" (-3=large negative effect, +3=large positive effect, 0=no effect)

Criteria	Rating		Concept 1	Concept 2	Concept 3	Concept 4	Concept 5	Concept 6	Concept 7	Concept 8
CTQ 1	4		3	2	3	-2	3			
CTQ 2	1		2	1	-2	2	2			
CTQ 3	3		1	0	1	1	1			
CTQ 4	5		3	-2	3	-1	3			
CTQ 5	3		2	2	2	2	2			
CTQ 6	4		1	1	1	-1	1			
CTQ 7	3		3	3	3	3	3			
CTQ 8	4		2	2	2	2	2			
CTQ 9	3		1	1	1	1	1			
CTQ 10	2		-1	-1	-1	-1	-1			
CTQ 11	4		-2	-2	-2	-2	-2			
CTQ 12	5		-3	-3	-3	-3	-3			
CTQ 13	2		1	1	1	1	1			
CTQ 14	5		-1	-1	2	-1	-1			
CTQ 15	5		-1	-1	-1	-1	-1			
CTQ 16	2		-1	3	-1	-1	-1			
CTQ 17	3		-1	-1	-1	-1	-1			
CTQ 18	1		-1	-1	-1	-1	-1			
note: insert rows above this line to maintain calculations										
Count of Positives Effects (>0)			10	9	10	7	10	0	0	0
Count of Negatives Effects (<0)			8	8	8	11	8	0	0	0
Count of No Effect (=0)			0	1	0	0	0	0	0	0
Weighted Sum of Positive Effects			64	47	72	33	64	0	0	0
Weighted Sum of Negative Effects			-41	-49	-38	-58	-41	0	0	0
Weighted Net Effect			**23**	**-2**	**34**	**-25**	**23**	**0**	**0**	**0**

FIGURE 14.4
Weighted Pugh Selection Matrix.

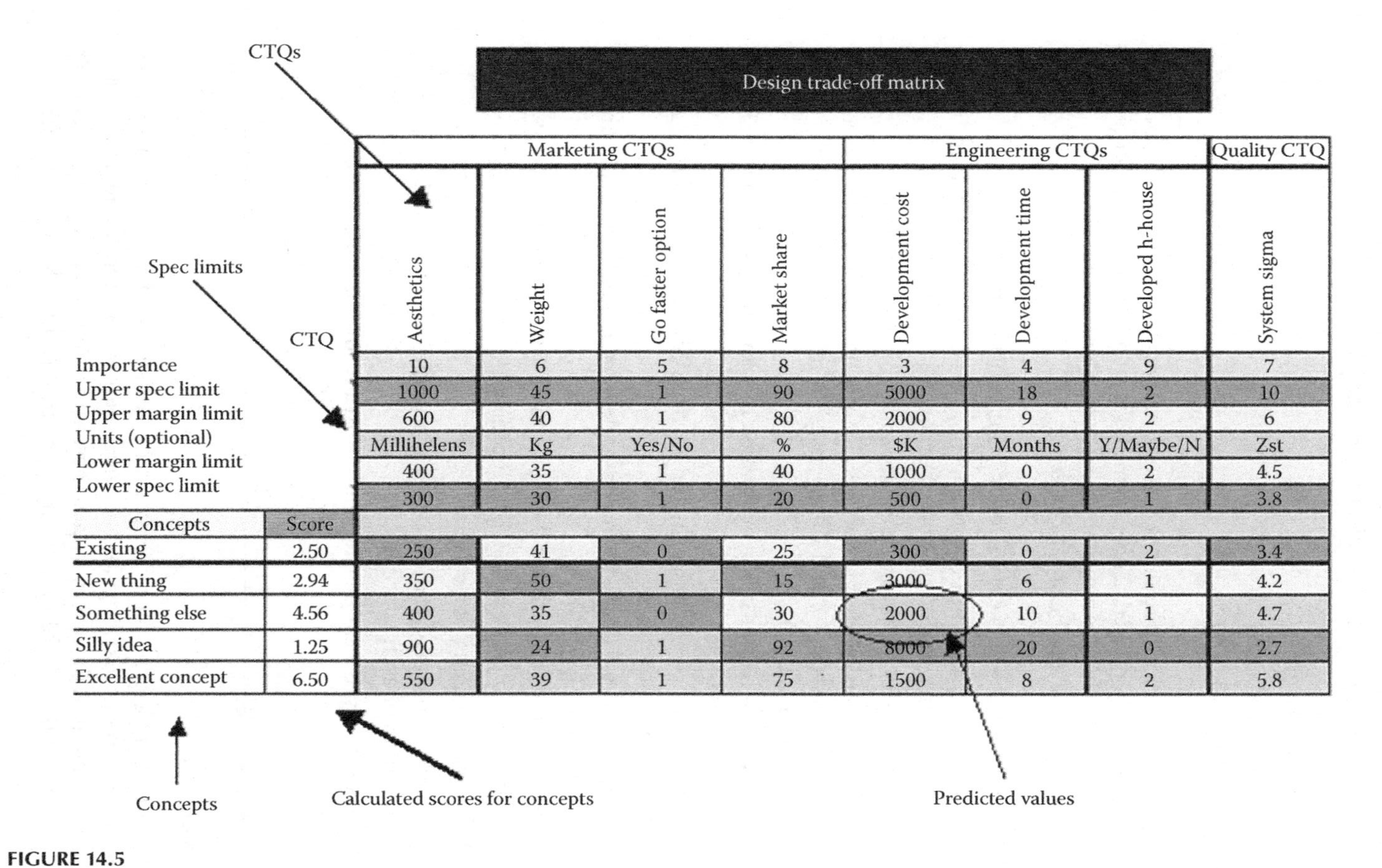

		Marketing CTQs				Engineering CTQs			Quality CTQ
	CTQ	Aesthetics	Weight	Go faster option	Market share	Development cost	Development time	Developed h-house	System sigma
Importance		10	6	5	8	3	4	9	7
Upper spec limit		1000	45	1	90	5000	18	2	10
Upper margin limit		600	40	1	80	2000	9	2	6
Units (optional)		Millihelens	Kg	Yes/No	%	$K	Months	Y/Maybe/N	Zst
Lower margin limit		400	35	1	40	1000	0	2	4.5
Lower spec limit		300	30	1	20	500	0	1	3.8
Concepts	Score								
Existing	2.50	250	41	0	25	300	0	2	3.4
New thing	2.94	350	50	1	15	3000	6	1	4.2
Something else	4.56	400	35	0	30	2000	10	1	4.7
Silly idea	1.25	900	24	1	92	8000	20	0	2.7
Excellent concept	6.50	550	39	1	75	1500	8	2	5.8

FIGURE 14.5
Design trade-off matrix.

concepts to an existing concept or standard, which is referred to as the "datum." The three main classification metrics include

- Same as datum
- Better than datum
- Worse than datum

The process typically involves several iterations to combine the best features of the highly ranked concepts until one superior concept emerges. This superior concept then becomes the new datum or benchmark.

The key steps for the PCSM include

1. Determine the concept selection criteria using the VOC.
2. Determine the best-in-class concept that should be the datum from industry rankings or customer surveys.
3. Develop six to eight candidate concepts for evaluation.
4. Evaluate each candidate concept against the datum using the selection criteria—(+), (-), and (S) are used to indicate better, worse, and same as the datum.
5. Analyze the results of the evaluation by summing the number of (+), (-), and (S) ratings.
6. Identify the weaknesses in the low-scoring concepts that can be turned into strengths.
7. Identify the weaknesses in the high-scoring concepts that can be turned into strengths.
8. Integrate the strengths of the concepts to create hybrid concepts.
9. Add any new concepts or hybrid concepts.
10. Reevaluate using Steps 4–8.
11. Based on the reevaluation, develop a single hybrid concept using the strengths of each concept.
12. If necessary, repeat the steps for a third round of evaluation.

Conclusion

Concepts can be generated from several sources such as personal inspiration, vision, benchmarking, and reverse engineering. Once a team develops multiple potential design concepts, these can be evaluated against customer criteria using the PCSM to determine the best candidate concept. This final concept can then be modeled to increase robustness. Modeling of technology will be discussed in Chapter 15.

Questions

1. How can design concepts be generated?
2. What methodology should be used to evaluate the set of candidate concepts?
3. What are the main steps in the Pugh Matrix?
4. How should design trade-offs be evaluated?
5. What criteria should be used in the Pugh Matrix?
6. What is the "datum"?
7. What are the classification metrics used in the PCSM?
8. What is a hybrid concept?
9. How is the final superior concept selected?

15

Modeling of Technology

The first rule of any technology used in a business is that automation applied to an efficient operation will magnify the efficiency. The second is that automation applied to an inefficient operation will magnify the inefficiency.

Bill Gates

Design for Six Sigma (DFSS) is a powerful tool for developing complex systems including products, services, and processes. To fully understand these systems, a key step within DFSS is modeling the relationships within the system architectures. Several tools that enable the DFSS team to model these technologies include the ideal function, P-diagram, and functional analysis system technique.

Ideal Function

The ideal function is used to represent the system input/output relationship. More specifically, it defines the desired relationship between the inputs and the outputs of a system, subsystem, product, or service. The inputs involve the various engineering control factors. The output is the measured critical functional response variable. Equation 15.1 represents the mathematical model for the customer-focused response if no noises are acting on the design or process.

$$Y = f(x) \tag{15.1}$$

The goal of DFSS is to develop a robust design. A design is said to be robust when the product or process is insensitive to the effects of variability without actually removing the sources of variability. Figure 15.1 represents the ideal function when there are no noises acting on the system.

Two key questions should be asked when using the ideal function:

1. How would we like the product to perform in the absence of noise?
2. What is the ideal relationship between the signal (customer input) and the quality (performance) characteristic?

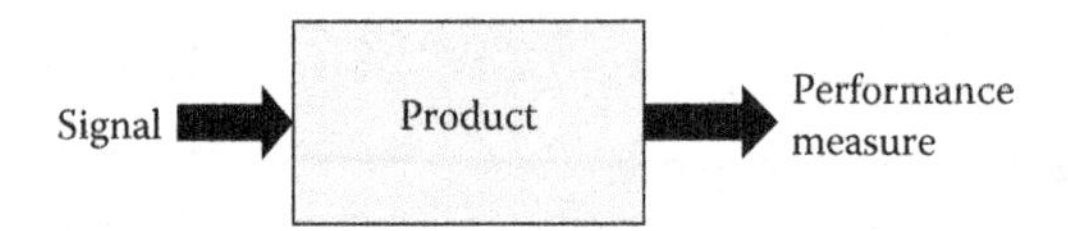

FIGURE 15.1
Ideal function.

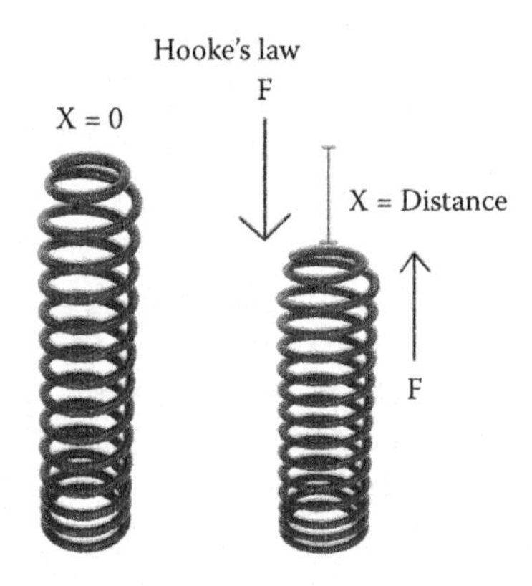

FIGURE 15.2
Helical compression spring.

An example of the ideal function can be shown through a helical compression spring (shown in Figure 15.2). Hooke's law provides the ideal function, which is $F = kx$ (shown in Figure 15.3), where

F = force
k = spring constant
x = displacement

The ideal function can then be exploited to provide the desired function. Side effects are anything other than the intended result. For a system to be robust, we need to understand the harmful side effects. The actual system performs under the mathematical model shown in Equation 15.2:

$$F = kx + E(x) \tag{15.2}$$

where $E(x) = $ error.

The function now takes into account the error, which is added to the useful or desired response. In other words, the function is the sum of the useful and the harmful part. This error is indicated in the variation around the function as shown in Figure 15.4.

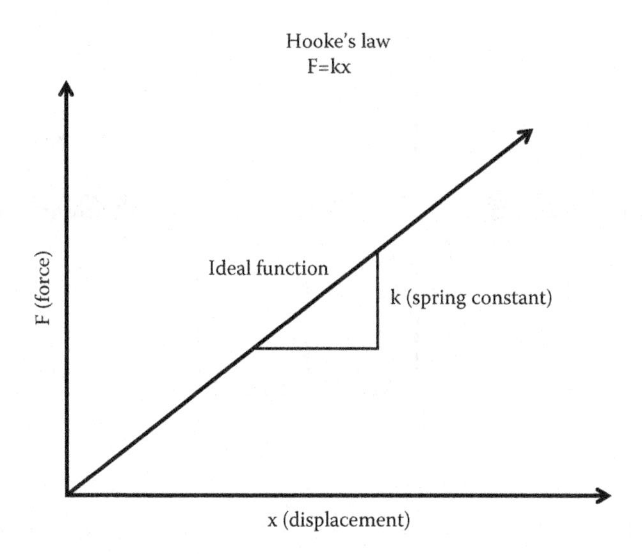

FIGURE 15.3
Ideal function for a helical compression spring.

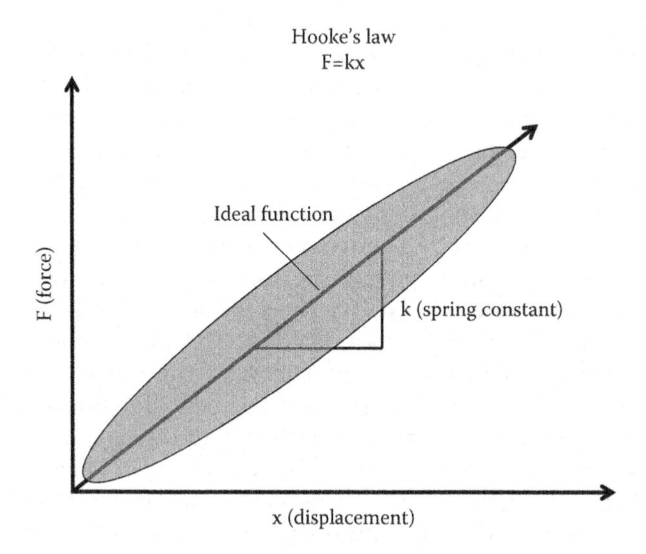

FIGURE 15.4
Actual performance for a helical compression spring.

P-Diagram

The ideal function enables a DFSS team to understand the desired function of a system without the presence of noise. It reflects the intended output necessary to meet the customer's expectations. The Parameter Diagram (P-diagram)

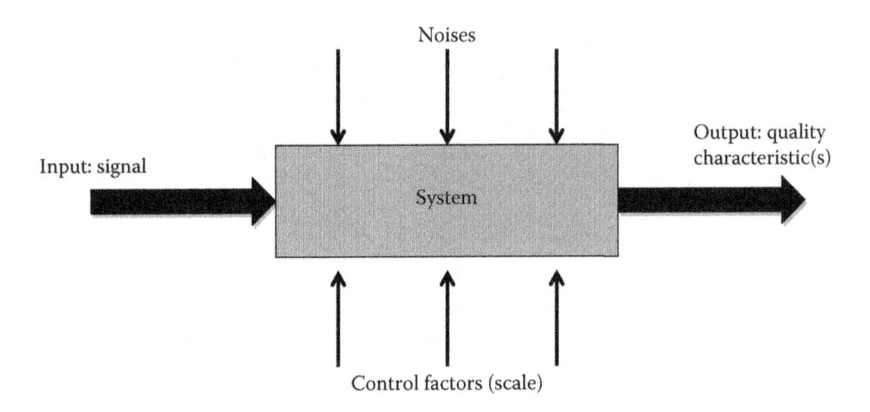

FIGURE 15.5
P-diagram.

is an extension of the ideal function, which then allows the DFSS team to understand and characterize the noises and control factors that impact a system. The P-diagram provides a graphical means to illustrate the input–output relationship of all parameters that can impact performance. When creating the P-diagram as a team, it is essential to use systems thinking to consider all of the possible parameters. Figure 15.5 illustrates a P-diagram.

The P-diagram contains information on the quality characteristic(s), control factors, noises, and signal. The first and most critical step is for the team to determine the quality characteristic(s), since this is how the system will be measured by the customer. This is the output of the system that must meet the customer's expectations as indicated by the voice of the customer. Therefore, the quality characteristic must be a measurable output, preferably a quantitative measure. The ideal function is a tool used to determine the useful output. However, the actual performance of a system contains the useful output and harmful side effects. The useful output is what should be measured, rather than the harmful side effects. When the output is maximized in terms of the useful output, the impact of the harmful side effects will be reduced. Therefore, it is critical that the DFSS team selects the appropriate quality characteristic as the performance measure.

The signal factor is the input to the system and is usually provided by the customer. When designing a system using DFSS, the DFSS team must understand and consider the reasonable range of this input factor.

Control factors are those parameters within the system that can be used to control the performance of the system. Control factors should be easy to control through minor adjustments by the user or the producer. For example, control factors could be specifications on product drawings. It is also important that control factors be independent to ensure that a change in one control factor does not impact the system performance in another area of the system, for example, a subsystem. Finally, since control factors can be used to change the performance of a system, these factors should have a large effect on the system performance.

Scale factors are a special type of control factor. Scale factors should have a large effect on the mean performance and a small effect on variation around the mean performance. In addition, scale factors should be easy to adjust. This enables the DFSS team to significantly impact the system performance through adjustments without affecting the overall variation in the system.

Noise factors are the sources of variation acting on the system. Noise factors are those parameters that either cannot be controlled or are too expensive to control: for example, humidity and temperature in a manufacturing facility. The DFSS team should determine the noise factors that have the greatest impact on the system performance. These will be the factors that the team will need to design the system to be robust enough to handle.

Functional Analysis System Technique

The functional analysis system technique (FAST) is a special type of tree diagram that is used to break down the high-level system functions to the lower-level functions. FAST is a technique that enables the DFSS team to graphically show the functional flow from the system to the subsystem and then from the subsystem to the parts. The FAST diagram enables a common understanding among the team members of how the system functions and the key relationships in the system. This also aids the team in considering various hardware solution sets. The inverse of the FAST diagram is the fault tree, which is commonly used to mitigate risk from system and product failures.

The FAST diagram provides a function tree that guides the DFSS team through the system design. The main purpose of the FAST diagram in DFSS is for the team to understand the overall system and the functional relationship between the systems, subsystems, subassemblies, and parts. Figure 15.6 illustrates the construction of a FAST diagram.

In developing a FAST diagram, an action verb + measurable noun function is used, in which the focus is on the positive function, for example, "deposit material." The use of the verb + noun structure places the focus of the DFSS team on visualizing the physical equations and transfer functions for each step in the FAST diagram. The requirements for the verb + noun statement are provided in the quality function deployment matrices. This allows the team to ensure that the detailed voice of the customer is considered in the functional design.

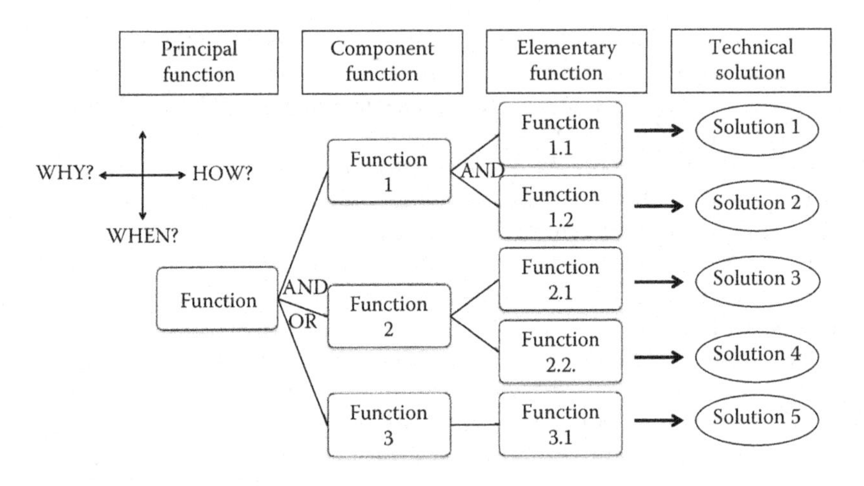

FIGURE 15.6
FAST diagram construction.

Conclusion

The FAST diagram should be used in conjunction with the fault tree and failure mode and effects analysis (FMEA), since those tools help the DFSS team to quantify and improve the robustness and reliability of the overall system. The P-diagram enables the team to define the input signal, noise and control factors, and quality characteristics. The FAST diagram is useful for providing the structure of the relationship of the systems architecture. It is critical to use these tools together to take the voice of the customer into the various methodologies for modeling the system. The next chapter on Taguchi Design discusses methodologies for improving the robustness of a product, process, or service.

Questions

1. What is the ideal function?
2. Give an example of an ideal function.
3. What function is used to express the correlations and the magnitude and directionality of sensitivity between system-level critical functional responses?
4. What type of variation is due to wear or some other form of degraded state or condition within the design?
5. What is a signal factor? Give an example.
6. What is a noise factor?

7. Describe the three categories of noise. Provide an example for each.
8. What calculation is used to relate the useful part of the response to the nonuseful variation in the response?
9. How does the P-diagram require systems thinking?
10. What does the P-diagram show?
11. What should the quality characteristic measure?
12. Should control factors be independent or dependent? Why?
13. How does the FAST diagram guide the quantification of robustness and reliability?
14. How is the FAST diagram related to a fault tree?

16

Taguchi Design

Cost is more important than quality but quality is the best way to reduce cost.

Genichi Taguchi

Taguchi Loss Function

The quadratic loss function (QLF), also known as the *quality loss function,* is a metric developed by Genichi Taguchi that focuses on achieving the target value rather than on performance within the wider specification limits. Genichi Taguchi, a Japanese engineer, developed the quality loss function based on the economic consequences of not meeting target specifications. Using the QLF allows continuous improvement teams to quantify improvement opportunities in monetary terms, the language of upper management. The QLF translates variability into economic terms by calculating the relationship between performance and financial outcome. The general QLF is shown in Equation 16.1.

$$\text{Loss at any point }(\text{L}) = (\text{monetary constant}) * (\text{average} - \text{target})^2 \quad (16.1)$$

The QLF is used to determine the average loss per product or encounter, and it enables Six Sigma teams to focus on performance relative to target and avoid the goalpost mentality. The loss function approximates the long-term loss from performance failures and encourages continuous improvement. The QLF is helpful both as a philosophical approach and as a quantitative method.

Taguchi's loss function combines cost, target, and variation into one metric, with specifications being of secondary importance. Losses of concern are those caused by a product's critical performance characteristic deviating from the target. The loss function is used to quantify the loss (quality cost) based on the deviation from the target. Taguchi also developed the concept of robustness, which means that noise factors are taken into account to ensure that the system functions correctly. The Taguchi loss function is the driving force behind continual process improvement. Figure 16.1 illustrates the QLF.

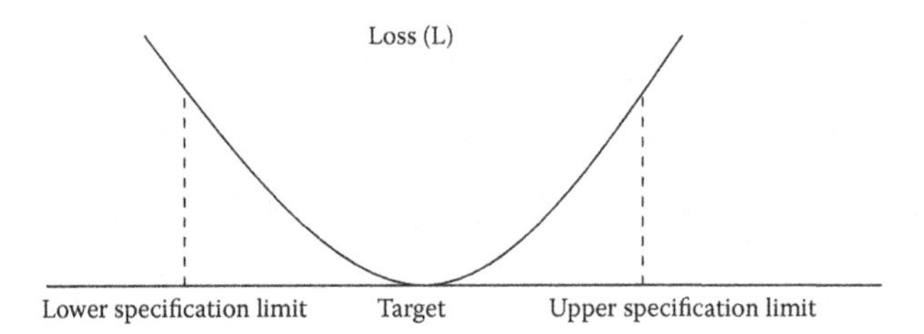

FIGURE 16.1
Taguchi loss function.

Taguchi defined quality as the "[avoidance of] loss a product causes to society after being shipped, other than any losses caused by its intrinsic functions" (Evans and Lindsay, 1999). Losses to society encompass costs incurred by the manufacturer, the customer, and anyone affected by product deficiencies.

The quality loss function unifies quality and cost to drive on-target engineering. It also relates economic and engineering terms in one model. The function enables optimization of costs through the production and use of a product (Fowlkes and Creveling, 1995).

The loss to society includes costs incurred due to a product or service not meeting customer expectations, not meeting performance characteristics, and having harmful side effects. The quality loss function associates a dollar value to the current state of quality for a product or process. The quality loss function approximates losses due to scrap, rework, poor performance, lack of customer satisfaction, and so on. This dollar amount can be used to identify areas for improvement and evaluate improvement efforts.

Taguchi relates the monetary loss to quantifiable product or service characteristics. This translates the language of an engineer into management terms. The quality loss function focuses on target values rather than specifications for process output. When nominal specifications are achieved, the products are consistent, and there is no loss to society (Breyfogle, 2003). Any deviation from the target incurs a loss to society. The loss increases as the measured response moves further away from the target value. The measured loss, $L(y)$, for a single product is estimated as shown in Equation 16.2.

$$L(y) = k(y - T)^2 \tag{16.2}$$

where:

- $L(y)$ is the loss in dollars
- y is the measured response
- T is the target value of the product's response
- k is the quality loss coefficient

The quality loss coefficient, k, is determined using the customer tolerance and economic consequence (Equation 16.3). The economic consequence, A_0, is the cost of replacing or repairing the product. The associated costs include losses incurred by the manufacturer, the customer, or a third party. The customer tolerance is the point at which the product reaches unacceptable performance.

$$k = \frac{A_0}{\Delta_0^2} \tag{16.3}$$

where:

A_0 is the economic consequence of failure

Δ_0 is the functional limit or customer tolerance for the measured response

The quality loss function can also be used to determine the average loss per product. The expected loss, $E(L)$, is used to depict the average loss. To reduce the estimated loss, the variability and deviation from the target must be reduced (Kiemele et al., 1999). Equation 16.4 shows the estimated loss.

$$E(L) = k\left(\sigma_y^2 + \left(\bar{y} - T\right)^2\right) \tag{16.4}$$

where:

σ_y^2 is the variation

$\bar{y}$ is the average response

T is the target value

k is the quality loss coefficient

Mahalanobis–Taguchi System

Prasanta Chandra Mahalanobis introduced the Mahalanobis distance (MD) in 1936. The Mahalanobis–Taguchi System (MTS) was later developed by Genichi Taguchi as a diagnosis and forecasting method using multivariate data for robust engineering.

MD is a distance measure that is based on correlations between variables and the different patterns that can be identified and analyzed with respect to a reference population. MD is a discriminant analysis tool (Taguchi and Jugulum, 2002), which is used to predict changes in output corresponding to changes in multiple engineering characteristics at all levels of a multidimensional system. MD is different from Euclidean distance (ED) because it addresses the correlations or distribution of the data points. Traditionally, the MD methodology has been used to classify observations into different groups.

ED is typically used to calculate the straight-line distance between two data points. ED can also be used to calculate the distance of an unknown point from the group mean. However, ED has two main disadvantages. First, ED does not determine a statistical measurement for how the unknown point matches the reference set. Second, ED only provides the relative distance from the mean point in the group. It does not account for the distribution of points in a group.

MD is a measure based on correlations between variables and the different patterns that can be identified and analyzed with respect to a reference point. MD is very useful in determining the similarity of a set of values to an unknown by comparing a sample from the unknown group with a measured collection of known samples. MD is different from ED because it addresses the correlations or distribution of the points, while ED is typically used to calculate the straight-line distance between two points. In a two-dimensional plane, the variables are defined as

$$p_1 = (x_1, x_1) \tag{16.5}$$

$$p_2 = (x_2, y_2) \tag{16.6}$$

where their ED is given by

$$d = \sqrt{(x_1 - x_2)^2 + (y_1 - y_2)^2} \tag{16.7}$$

Traditionally, the MD methodology has been used to classify observations into different groups. MD is defined in Equation 16.8.

$$\text{MD}_j = D_j^2 = \frac{1}{k} Z_{ij}^T A^{-1} Z_{ij} \tag{16.8}$$

where:

k = total number of variables
i = number of variables ($i=1, 2,\ldots, k$)
j = number of samples ($j=1, 2,\ldots, n$)
Z_{ij} = standardized vector of normalized characteristics of x_{ij}
= $x_{ij}-m_i/s_i$
x_{ij} = value of the ith characteristic in the jth observation
m_i = mean of the ith characteristic
s_i = standard deviation of the ith characteristic
T = transpose of the vector
A^{-1} = inverse of the correlation matrix

MD is used to determine the similarity of a known set of values to that of an unknown set of values. MD has been successfully applied to a broad range of cases, mainly because it is very sensitive to intervariable changes in data. Also, since the MD is measured in terms of standard deviations from the mean of the samples, it provides a statistical measure of how well an unknown sample matches a known sample set.

For example, consider a data set of different samples from the same material. When measured, the samples of the same material will be very similar; however, no two samples will be exactly alike. There will be slight differences due to drift, handling, environmental factors, and batch-to-batch variation. Since the samples are all from the same material, the relative intensities of the attributes will remain approximately the same. If different samples of the sample material are measured, and two attributes are plotted with the responses for the first attribute versus the responses for the second attribute, the plot will resemble Figure 16.2.

MD is a discriminant analysis tool that is used in this research to translate the lower-level functions into an estimate of consumer satisfaction at the product level. MD is used to predict how changes in engineering characteristics in the product design impact consumer satisfaction. Taguchi (2000) developed a method for calculating the MD using eight major steps, as shown in Table 16.1.

MD methodology differs from other classical statistical approaches such as cluster analysis, principal components analysis, and factor analysis. First, MD considers the variance and covariance of the measured variables rather than just the average value. It weights the differences by the variability range in the sample point direction. This accounts for natural variation within a set. Second, it also accounts for the ranges of acceptability between the variables. It compensates for the interactions between variables. This is useful, since most systems are not composed of independent variables, but rather, systems of dependent variables impacting the system performance. Also, it calculates distances in units of standard deviation from the group mean. By use of the standard deviation, the attribute under consideration is given in the original units, which relates the results to the original data. The calculated

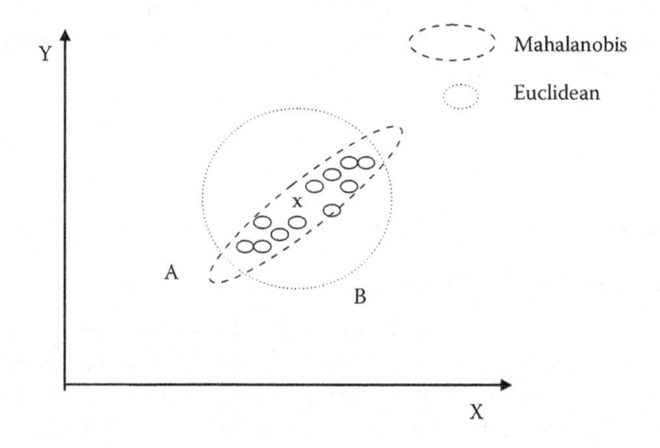

FIGURE 16.2
Representation of ED and MD.

TABLE 16.1

Steps for Calculating MD

1. Define the normal group
2. Determine k variables or characteristics
    a. 2⟨k⟨1000
3. Gather data on the variables from the normal group using sample size of n
    a. k⟨⟨n
4. Calculate the MD D^2 for each sample
    a. Normalize the data
    b. Calculate the correlation matrix
    c. Invert the correlation matrix
    d. Calculate the MD for each of the n samples using the inverted correlation matrix
5. Gather data on the variables from outside of the reference group using a sample size of r
6. Calculate the MD D^2 for each sample from the test group
7. Evaluate the discrimination power by verifying that there is a difference between the normal and test groups
8. Optimize the system
    a. Assign the variables to a two-level orthogonal array (OA)
        i. Define use = 1, do not use = 2
        ii. Each row of the OA represents a subset of the variables for the MD calculation
    b. Assign the best estimate of true value for each sample in the test group. Estimated values are the input signal values
    c. Calculate the MD for the nonreference samples in each row of the OA
    d. Calculate the dynamic signal-to-noise ratio for each row of the OA
    e. Eliminate the variables that do not significantly influence the S/N ratio
    f. Conduct a confirmation run with the remaining variables
    g. Evaluate the discrimination power by verifying that there is a difference between the normal and test groups

circumscribing ellipse around a cluster defines a one-standard-deviation boundary of that group. Normally distributed variables can be converted into probabilities using the χ^2 (Chi-square) probability density function. The probability of a group being normal or abnormal can be calculated for a confidence interval.

Multidimensional Systems

A system, by its nature, is multidimensional. To improve or optimize a system, it is necessary to understand the variables and noise conditions that affect system performance. Not all variables affect system performance to an equal degree. Therefore, it is essential to identify the critical sets of variables through discriminant analysis of the system. The critical sets of variables can then be used for diagnosis and prediction.

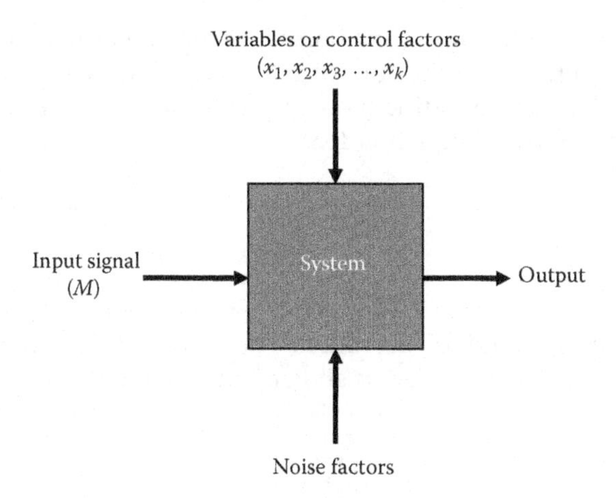

FIGURE 16.3
Multidimensional system.

A multidimensional system consists of an input signal, *M*, which is typically given by the consumer. Noise conditions arise from changes in consumer use of the product, wear and deterioration, and manufacturing variation. The control factor set contains those factors or design variables that are used to modulate the performance of the system, product, or service. The combination of all of the levels of the control factors contains the desired levels. The control factor level set spans the feasible design space. The output, *y*, should ideally provide fidelity to the input signal and be a measurable characteristic. A graphical depiction of a multidimensional system is given in Figure 16.3.

The MTS can be used to minimize the number of variables required for diagnosis and to predict the performance of the system.

Mahalanobis–Taguchi Steps

Step 1: Normal Group Calculations

The first step in performing MTS is to define a "normal" or "healthy" group, which defines the Mahalanobis space (MS). The normal group is selected with discretion to define a reference point on a measurement scale. Defining the normal group is a critical step in this method, since the MS is the reference point and base of the measurement scale. An abnormal condition is outside the healthy group. The degree of abnormality is measured with reference to the normal group.

1. Calculate the mean. Once the normal group is selected, the mean for each characteristic in the normal group is calculated by summing all of the values and dividing by the number of samples in the normal group, as shown in Equation 16.9.

$$\overline{x} = \frac{\sum_{i=1}^{N} x_i}{N} \tag{16.9}$$

2. Calculate the standard deviation. The standard deviation for each characteristic is calculated as shown in Equation 16.10.

$$s = \sqrt{\frac{\sum_{i=1}^{N} (x_i - \overline{x})^2}{N}} \tag{16.10}$$

   MTS considers the normal and abnormal groups as separate populations; therefore, the standard deviation calculation uses N in the denominator rather than N–1.

3. Normalize the data. Each characteristic is then normalized using the mean and standard deviation values.

$$Z_i = \frac{x_i - \overline{x}}{s} \tag{16.11}$$

   When the values are normalized, the MS of the normal group has a mean at the zero point and an average MD of unity. Since the average MD for the normal group is unity, it is also called the *unit space* or the MS.

4. Calculate the mean of the normalized data. The mean of the normalized data is next calculated using the standard mean equation, as shown in Equation 16.12.

$$\overline{Z} = \frac{\sum_{i=1}^{N} Z_i}{N} \tag{16.12}$$

5. Calculate the standard deviation of the normalized data. The standard deviation of the normalized data is then given by Equation 16.13.

$$s = \sqrt{\frac{\sum_{i=1}^{N} \left(Z_i - \overline{Z}\right)^2}{N}} \tag{16.13}$$

In MTS, the mean of the standard deviations for the normalized data will always equal one.

6. Calculate the correlation matrix for normalized values. Next, the correlation matrix is calculated using the normalized values of the characteristics. The correlation matrix calculates the correlation coefficients between the variables.

$$r = \frac{\sum Z_1 Z_2}{N} \tag{16.14}$$

7. Calculate the inverse of the correlation matrix. The inverse matrix exists only for a nonsingular matrix. The inverse matrix B of matrix A satisfies the equation AB = BA = I. The inverse of matrix A is denoted as A^{-1}.
8. Calculate the matrix product AA^{-1}. The product of the matrix multiplication of AA^{-1} is a validation of numerical accuracy. The product should give the identity matrix.
9. Calculate *W* values. The *W* values are calculated by multiplying the matrix of normalized data by the inverse matrix (Equation 16.15).

$$W_i = Z_{\text{Normal}} A^{-1} \tag{16.15}$$

Step 2: Normal Space Calculations

The MD is then calculated for the normal group. MTS introduces a measurement scale of all input characteristics to measure the degree of abnormality. The MD is scaled by dividing by the number of variables:

$$D^2 = \frac{W_1 Z_1 + W_2 Z_2 + W_3 Z_3 + \cdots + W_i Z_i}{i} \tag{16.16}$$

Step 3: Test (Abnormal) Group Calculations

1. Calculate the *Z* values for the test group. The *Z* values for the test group, otherwise known as the abnormal group, are calculated using the mean and standard deviation from the normal group:

$$Z_i = \frac{x_i - \overline{x}}{s} \tag{16.17}$$

2. Calculate the *W* values for the test group. The *W* values for the test group are calculated by multiplying the matrix of normalized

data from the test group by the inverse matrix from the normal group.

$$W_i = Z_{Test}A^{-1} \tag{16.18}$$

3. Calculate MD (D^2) for the test group. The MD is calculated for the test group by summing the multiplication of the W values by the normalized values and dividing by the number of characteristics:

$$D^2 = \frac{W_1Z_1 + W_2Z_2 + W_3Z_3 + \cdots + W_iZ_i}{i} \tag{16.19}$$

Step 4: Optimize the System

1. Choose the orthogonal array (OA). In this step, the purpose is to reduce the dimensionality of a multivariate system and obtain meaningful results. OAs are used to estimate the effect of factors and interactions of factors by minimizing the number of experiments required. The appropriate OA is selected based on the number of factors and the number of levels. An OA is employed in the MTS to minimize the number of runs in an experiment.
2. Calculate the dynamic signal-to-noise (S/N) ratio. For each experiment in the OA, an S/N ratio is calculated. S/N ratios measure the magnitude of the signals in a system. The S/N ratio measures the functionality of a system and the interaction between control factors and noise factors. There are two types of S/N ratios: static and dynamic. The static S/N ratio is used when the signal factor is absent or has a fixed value. The dynamic S/N ratio is used to determine a system's response to specific levels of the signal factor. Dynamic S/N is used when the signal and response must follow a function. Dynamic S/N ratios are useful for technology development. The equation for the dynamic S/N ratio is given in Equation 16.20 (Taguchi, 2002).

$$\frac{S}{N} = 10\log_{10}\frac{\frac{1}{r}\left(S_\beta - V_e\right)}{V_e} \tag{16.20}$$

where:
S_T = total sum of squares

$$S_T = \sum_{i=1}^{t} y_i^2 \tag{16.21}$$

r = sum of squares due to input signal (power of the input signal)

$$r = 2\sum_{i=1}^{t} M_i^2 \tag{16.22}$$

S_β = sum of squares due to slope

$$S_\beta = \frac{1}{r}\sum_{i=1}^{t}(M_i y_i)^2 \tag{16.23}$$

S_e = error sum of squares

$$S_e = S_T - S_\beta \tag{16.24}$$

V_e = error variance

$$V_e = \frac{S_e}{t-1} \tag{16.25}$$

3. Determine the factor effects using the dynamic S/N ratio. Based on the S/N ratios calculated using the runs in the OA, average response tables are constructed to determine the useful variables and identify the candidate variables for elimination. For S/N ratios, the larger the number in the positive direction, the greater the impact of the characteristic on the system.

MTS Steps Using the Graduate Admission System Example

The graduate admission system MTS example presented in *The Mahalanobis–Taguchi Strategy* (Taguchi, 2002) is used to illustrate the MTS. This example was selected because the size of the problem is relatively small, and numerous publications exist. Also, the first step in fully understanding the calculations involved in MTS was to perform all of the calculations by hand. After the hand calculations were performed, a spreadsheet was developed to calculate this example.

The graduate admission system consists of three variables: grade point average (GPA) (X1), SAT (X2), and SAT-Math (X3). These characteristics and their designated variable notations are shown in Table 16.2.

The normal group consists of graduate admission scores for successful graduates. The normal group was defined with 15 samples ($n=15$) and 3 variables ($k=3$), as shown in Table 16.3.

TABLE 16.2

Graduate Admission System Variables

Variable	Characteristic
X1	GPA
X2	SAT
X3	SAT-Math

TABLE 16.3

Normal Group Data

	GPA	SAT	SAT-Math
	X1	X2	X3
1	3.0	1010	670
2	2.9	990	428
3	3.2	1035	712
4	2.8	980	546
5	3.9	1310	677
6	3.2	990	650
7	2.8	965	646
8	3.7	1380	715
9	3.4	1300	645
10	3.2	1205	645
11	2.6	895	490
12	3.1	950	555
13	3.6	1110	520
14	2.7	1045	625
15	3.7	1235	690
Mean	3.19	1093.33	614.27
Std. Dev.	0.39	147.85	83.25

Step 1: Normal Group Calculations

1. Calculate the mean. The mean for each characteristic in the normal group of the graduate admission system example is calculated as

$$\bar{x} = \frac{\sum_{i=1}^{N} x_i}{N}$$

$$\bar{x} = 3.18667$$

2. Calculate the standard deviation. The standard deviation for each characteristic in the normal group is calculated. In the literature for

this example, the standard deviation calculation for a population is used, although the number of samples is relatively small. This example will also use the standard deviation equation for a population, to remain consistent with the literature.

$$s = \sqrt{\frac{\sum_{i=1}^{N}(x_i - \bar{x})^2}{N}}$$

$$s = \sqrt{\frac{2.25734}{15}}$$

$$s = 0.38793$$

3. Normalize the data. The next step in the MTS is to normalize the data from the normal group.

$$Z_i = \frac{x_i - \bar{x}}{s}$$

$$Z_1 = \frac{3.0 - 3.18667}{0.38793}$$

$$Z_1 = -0.48120$$

4. Calculate the mean of the normalized data. Now that the data is normalized, the mean for each characteristic in the normal group is now the zero point. This is illustrated in Table 16.4, where the mean for all normal group characteristics is equal to zero.

$$\bar{Z} = \frac{\sum_{i=1}^{N} Z_i}{N}$$

$$\bar{Z}_1 = \frac{-0.00014}{15}$$

$$\bar{Z}_1 = 0.000$$

5. Calculate the standard deviation of the normalized data. The standard deviation of the normalized data for each characteristic in the normal group is one. This is shown in Table 16.4, in which the standard deviation for all characteristics is equal to 1.0.

TABLE 16.4

Normalized Data from the Normal Group

	GPA	SAT	SAT-Math
	Z1	Z2	Z3
1	−0.481	−0.564	0.669
2	−0.739	−0.699	−2.237
3	0.034	−0.395	1.174
4	−0.997	−0.767	−0.820
5	1.839	1.465	0.754
6	0.034	−0.699	0.429
7	−0.997	−0.868	0.381
8	1.323	1.939	1.210
9	0.550	1.398	0.369
10	0.034	0.755	0.369
11	−1.512	−1.341	−1.493
12	−0.223	−0.969	−0.712
13	1.065	0.113	−1.132
14	−1.255	−0.327	0.129
15	1.323	0.958	0.910
Mean	0.000	0.000	0.000
Std. Dev.	1.000	1.000	1.000

$$s = \sqrt{\frac{\sum_{i=1}^{N}\left(Z_i - \bar{Z}\right)^2}{N}}$$

$$s = \sqrt{\frac{14.970}{15}}$$

$$s = 1.000$$

6. Calculate the correlation matrix for the normalized values. The correlation matrix for the normalized values, which gives the correlation coefficient between the variables, is calculated in the following equations. Table 16.5 shows the correlation matrix.

$$r = \frac{\sum_{i=1}^{N} Z_1 Z_2}{N}$$

$$r(Z_1, Z_1) = 1.000$$

TABLE 16.5

Correlation Matrix

	Z1	Z2	Z3
Z1	1.000	0.832	0.485
Z2	0.832	1.000	0.557
Z3	0.485	0.557	1.000

7. Calculate the inverse of the correlation matrix. The next step in MTS is to calculate the inverse of the correlation matrix, as shown in Table 16.6.
8. Multiply AA^{-1}. The correlation matrix is multiplied by the inverse of the correlation matrix, resulting in an identity matrix. This is a validation step to confirm that all steps up to this point are correct. The multiplication of the correlation matrix by the inverse of the correlation matrix should result in the identity matrix. For two square matrices, B is said to be the inverse of A if AB = BA = I. If B is the inverse of A, it is denoted as A^{-1}. Table 16.7 shows the resulting identity matrix.
9. Calculate *W* values. The *W* values are calculated by multiplying the matrix of the normalized data by the inverse of the correlation matrix (A^{-1}). Table 16.8 shows the *W* values.

$$W_i = ZA^{-1}$$

$$W(1,1) = (-0.481)(3.254) + (-0.564)(-2.650) + (0.669)(-0.102)$$

$$W(1,1) = -0.140$$

TABLE 16.6

Inverse of the Correlation Matrix

	1	2	3
1	3.254	−2.650	−0.102
2	−2.650	3.609	−0.726
3	−0.102	−0.726	1.454

TABLE 16.7

Multiplication of AA^{-1}

	1	2	3
1	1.000	0.000	0.000
2	0.000	1.000	0.000
3	0.000	0.000	1.000

TABLE 16.8

Calculations of *W* Values for Normal Group

	W_1	W_2	W_3
1	−0.140	−1.245	1.431
2	−0.325	1.060	−2.670
3	1.038	−2.367	1.990
4	−1.128	0.470	−0.534
5	2.023	−0.131	−0.155
6	1.920	−2.925	1.128
7	−0.982	−0.768	1.286
8	−0.956	2.613	0.217
9	−1.953	3.320	−0.534
10	−1.927	2.367	−0.015
11	−1.214	0.250	−1.043
12	1.915	−2.390	−0.309
13	3.283	−1.595	−1.836
14	−3.229	2.051	0.552
15	1.674	−0.709	0.493

Step 2: Normal Space Calculations

MD is calculated for each sample in the normal group. The D^2 calculation treats this as a vector multiplication. Table 16.9 shows the MD values for the normal group.

$$D^2 = \frac{1}{k}\sum_{ij=1}^{k} a_{ij}\left(\frac{x_i - m_i}{\sigma_i}\right)\left(\frac{x_j - m_j}{\sigma_j}\right)$$

$$D^2 = \frac{1}{k}\sum_{ij=1}^{k} W_i\left(\frac{x_j - m_j}{\sigma_j}\right)$$

$$D^2 = \frac{1}{k}\sum_{ij=1}^{k} W_i Z_i$$

$$D_1^2 = \frac{(-0.140)(-0.481)+(-1.245)(-0.564)+(1.431)(0.669)}{3}$$

$$D_1^2 = 0.576$$

TABLE 16.9

Normal Group MD

	D^2
1	0.576
2	1.825
3	1.102
4	0.401
5	1.137
6	0.865
7	0.712
8	1.355
9	1.123
10	0.572
11	1.019
12	0.703
13	1.799
14	1.150
15	0.661
Average	1.000

TABLE 16.10

Graduate Admission System Data for Test Group

	GPA	SAT	SAT-Math
	X1	X2	X3
Clinton	2.40	1210	540
Monica	1.80	765	280
Paul	0.90	540	280
John	3.60	990	230
George	2.10	930	480
Ringo	2.60	1140	530
Genichi	4.00	1600	800

The test group is any group of data for which it is not known whether it is normal or abnormal. It is compared with the normal group MD to determine the degree of abnormality. The MD for the test group is given in Table 16.10.

Step 3: Test Group Calculations

1. Calculate the Z values for the test group (normalize the data). The Z values for the test group are calculated using the mean and standard

TABLE 16.11

Normalized Values for the Test Group

	Z1	Z2	Z3
Clinton	−2.028	0.789	−0.892
Monica	−3.575	−2.221	−4.015
Paul	−5.895	−3.743	−4.015
John	1.065	−0.699	−4.616
George	−2.801	−1.105	−1.613
Ringo	−1.512	0.316	−1.012
Genichi	2.097	3.427	2.231

deviation from the normal group. Table 16.11 shows the Z values for the test group.

$$Z_i = \frac{x_i - \overline{x}}{s}$$

$$Z_1 = \frac{2.40 - 3.18667}{0.38793}$$

$$Z_1 = -2.02787$$

2. Calculate the W values for the test group. The next step in MTS is to calculate the W values for the test group by multiplying the matrix of normalized data from the test group by the inverse of the correlation matrix from the normal group, as shown in Table 16.12.

$$W_i = ZA^{-1}$$

$$W_1 = \frac{(-2.028)(3.254) + (0.789)(-2.650) + (-0.892)(-0.102)}{3}$$

$$W_1 = -8.599$$

TABLE 16.12

W Values for Test Group

	W_1	W_2	W_3
Clinton	−8.599	8.869	−1.663
Monica	−5.337	4.372	−3.862
Paul	−8.853	5.028	−2.522
John	5.789	−1.996	−6.312
George	−6.023	4.607	−1.258
Ringo	−5.654	5.881	−1.547
Genichi	−2.487	5.193	0.543

3. Calculate MD (D^2) for the test group. D^2 is calculated for the test group by multiplying the W values by the normalized values. Tables 16.13 and 16.14 show the MD values for the normal and test groups, respectively.

$$D^2 = \frac{1}{k}\sum_{ij=1}^{k} a_{ij}\left(\frac{x_i - m_i}{\sigma_i}\right)\left(\frac{x_j - m_j}{\sigma_j}\right)$$

$$D^2 = \frac{1}{k}\sum_{ij=1}^{k} W\left(\frac{x_j - m_j}{\sigma_j}\right)$$

TABLE 16.13

Normal Data MD

	Normal
1	0.576
2	1.825
3	1.102
4	0.401
5	1.137
6	0.865
7	0.712
8	1.355
9	1.123
10	0.572
11	1.019
12	0.703
13	1.799
14	1.150
15	0.661

TABLE 16.14

Test Data MD

	Abnormal
Clinton	8.640
Monica	8.292
Paul	14.498
John	12.232
George	4.604
Ringo	3.991
Genichi	4.598

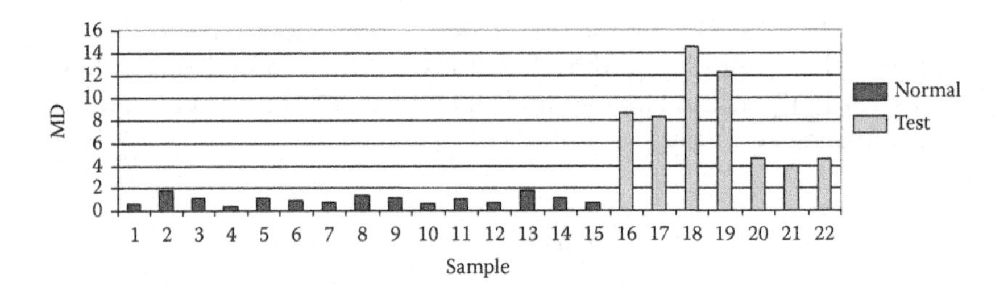

FIGURE 16.4
Normal group vs. test group graph.

$$D^2 = \frac{1}{k}\sum_{ij=1}^{k} WZ$$

$$D^2 = \frac{W_1Z_1 + W_2Z_2 + W_3Z_3 + \cdots + W_iZ_i}{i}$$

$$D^2_{\text{Clinton}} = \frac{(-8.599)(-2.028) + (8.869)(0.789) + (-1.663)(-0.892)}{3}$$

$$D^2_{\text{Clinton}} = 8.640$$

A comparison of normal group MD and test group MD values is given in Figure 16.4. The bar chart illustrates the significant difference in MD values between the normal and test groups.

Step 4: Optimize the System

1. Apply OA. The OA is selected based on the number of factors and the number of levels. In this case, an $L_8(2^3)$ OA was used to determine the number of runs. The general guidelines published (Fowlkes and Creveling, 1995) based on the number of factors and levels are used to determine the appropriate OA. A three-factor–two-level experiment is used to optimize the system following an L_8 OA.

   The three factors are GPA (X1), SAT (X2), and SAT-Math (X3). The two levels are include and exclude. Include means to use the factor to construct the space. Exclude means to ignore the factor and not use it to construct the space. This shows the effect of each factor on MD. For example, in the run (1, 1, 2), only the factors GPA and SAT are used. The third factor, SAT-Math, is not included in the MD calculation, and only two variables are considered.

2. Evaluation matrix. Because dynamic S/N ratios are used, an evaluation matrix is constructed to assign levels to the input signals. These input values or severity levels, M_i, are generally assigned arbitrarily. Larger values of M_i indicate a greater degree of abnormality. The M values are levels of the signal factor that span the range of expected values. The relationship between the input signals (M) and the output (y) is given by Equation 16.26 (Taguchi, 2002).

$$y_i = \beta M_i \tag{16.26}$$

where:

$$y_i = \sqrt{MD}$$

β is the slope
M_i is the working mean of the ith class

There must be at least three levels of the signal factors to show curvature or nonlinearity. The signal levels are chosen to span a reasonable range for the signal factors. Typically, one is near (but not on) the lower limit, one is at the target, and one is near (but not on) the upper limit. The M values selected in this case study are shown in Table 16.15.

An L_8 OA is used to optimize the system, as shown in Table 16.16.

3. Calculate the dynamic S/N ratio. The S/N ratio is calculated using the dynamic S/N equation as given by Equation 16.27.

$$\frac{S}{N} = 10\log_{10}\frac{\frac{1}{r}\left(S_\beta - V_e\right)}{V_e} \tag{16.27}$$

$$\frac{S}{N} = 10\log_{10}\frac{\frac{1}{2800}\left(537.27 - 0.62\right)}{0.62}$$

TABLE 16.15

Evaluation Matrix

	Point	Evaluation	Sample 1	Sample 2
M_1	10	Not so bad	George	Ringo
M_2	20	Poor	Clinton	Monica
M_3	30	Very poor	Paul	John

TABLE 16.16

L_8 OA

	A=X1	B=X2	C=X3	$M_1=10$		$M_2=20$		$M_3=30$	
	1	2	3	George	Ringo	Clinton	Monica	Paul	John
1	1	1	1	4.604	3.991	8.640	8.292	14.498	12.232
2	1	1	2	6.361	5.164	12.008	7.309	19.561	4.647
3	1	2	1	3.965	1.194	2.062	9.790	18.245	17.796
4	1	2	2	7.847	2.287	4.112	12.777	34.746	1.135
5	2	1	1	1.331	1.074	1.598	8.061	9.703	13.199
6	2	1	2	1.220	0.100	0.623	4.932	14.007	0.488
7	2	2	1	2.601	1.025	0.796	16.122	16.122	21.305
8	2	2	2	–	–	–	–	–	–

$$\frac{S}{N} = -5.10$$

where:

S_T = total sum of squares

$$S_T = \sum_{i=1}^{t} y_i^2 \quad (16.28)$$

r = sum of squares due to input signal

$$r = \sum_{i=1}^{t} M_i^2 \quad (16.29)$$

S_β = sum of squares due to slope

$$S_\beta = \frac{1}{r}\sum_{i=1}^{t} (M_i y_i)^2 \quad (16.30)$$

S_e = error sum of squares

$$S_e = S_T - S_\beta \quad (16.31)$$

V_e = error variance

$$V_e = \frac{S_e}{t-1} \quad (16.32)$$

4. Determine the factor effects using the dynamic S/N ratio. The next step in applying MTS is to determine the factor effects using the dynamic S/N ratio. This calculation is performed for the two factor levels. For example, the effect of a factor at the high level is calculated by adding the S/N ratios for that factor at the high level and dividing by the number of runs in which the factor is at the high level. Table 16.17 shows the factor effects.

$$A1 = \frac{\sum A1}{N1}$$

$$A1 = \frac{\sum(-5.104 - 21.54 - 19.20 - 27.10)}{4}$$

$$A1 = -18.24$$

The dynamic S/N ratio for factor effects can then be plotted to visually depict the magnitude of each factor's effect on the system. Correspondingly, the factor interaction effects can also be plotted, as shown in Figures 16.5 through 16.8. Figure 16.5 shows the gains for the three factors.

The next step is to express the functionality of the system in terms of S/N ratios using the gain calculation. The S/N ratio gain indicates the functionality improvement in the system measured in decibels. The gain is calculated by subtracting the S/N values for an attribute at Level 2 from those at Level 1 in the orthogonal array. Using the dynamic S/N ratio, all characteristics have positive gains for the graduate admissions example, as shown in Figure 16.9. Therefore, it is determined that it is best to include all three characteristics.

TABLE 16.17

Factor Effects

	S/N		B1
A1	−18.24	A1	−13.32
A2	−25.93	A2	−24.04
			C1
B1	−18.68	A1	−12.15
B2	−25.49	A2	−20.45
			C1
C1	−16.30	B1	−12.39
C2	−27.87	B2	−20.22

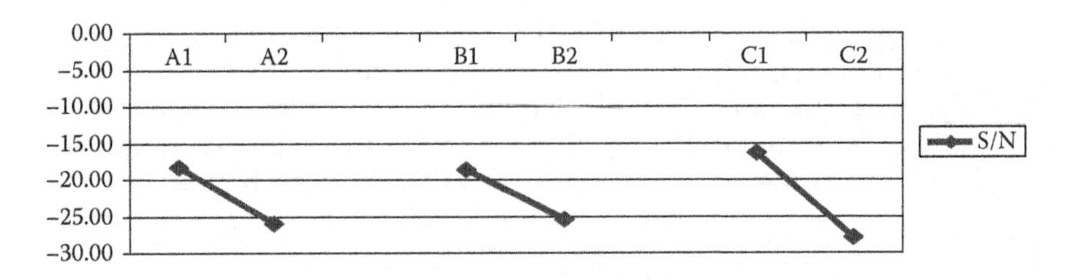

FIGURE 16.5
Dynamic signal to noise.

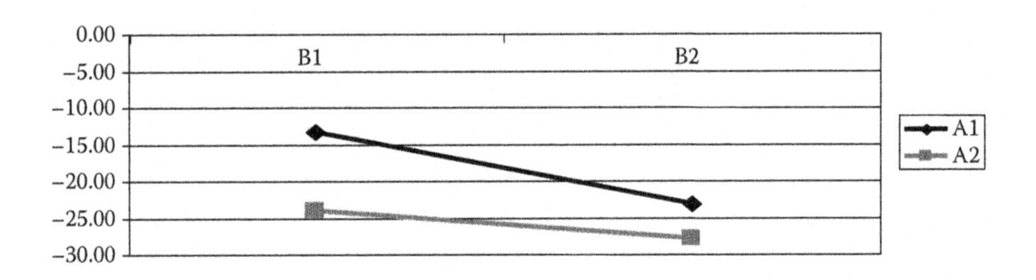

FIGURE 16.6
A–B interaction plot.

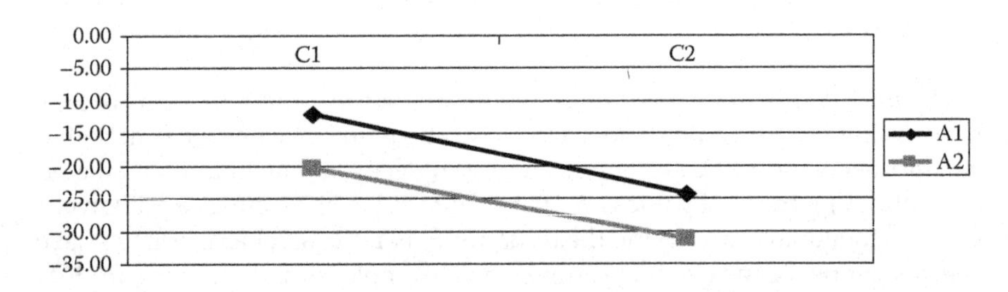

FIGURE 16.7
A–C interaction plot.

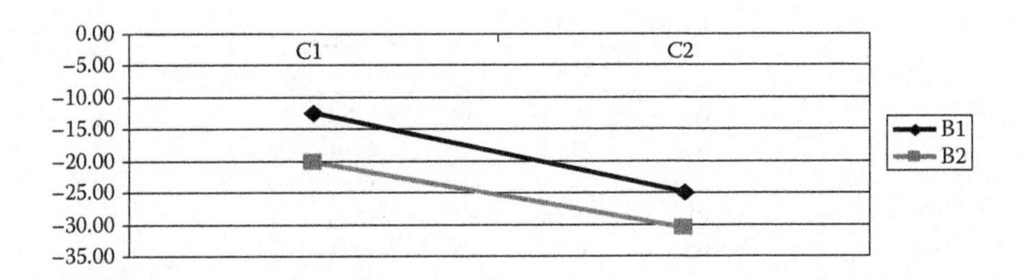

FIGURE 16.8
B–C interaction plot.

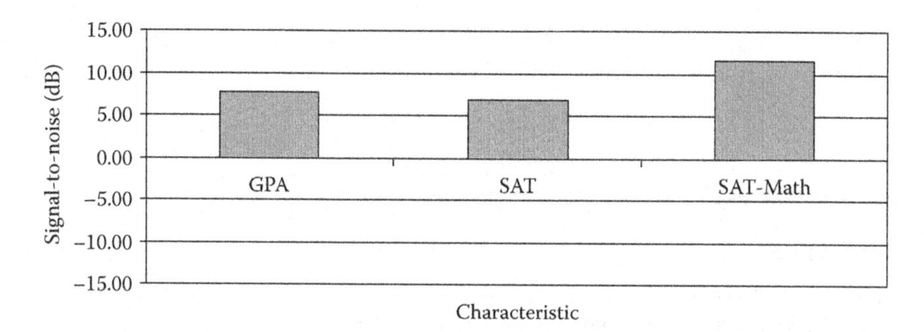

FIGURE 16.9
Gains.

Conclusion

The quality loss function is a metric of the economic consequences of not meeting target specifications that allows continuous improvement teams to quantify improvement opportunities in monetary terms. It translates variability into economic terms by calculating the relationship between performance and financial outcome. The MTS is a diagnostic and forecasting method using multivariate data for robust engineering that can be used to minimize the number of variables required for diagnosis and to predict the performance of the system. Chapter 17 presents a risk analysis tool known as the Design Failure Modes and Effects Analysis, which further improves the reliability of a product, process, or service.

Questions

1. What methodology is used to translate variability into economic terms by calculating the relationship between performance and financial outcome?
2. Why is the "goalpost mentality" considered bad?
3. What is robustness?
4. Which quality pioneer suggests that a reduction in the variation of a product or process represents a lower loss to society?
5. Why is the Taguchi loss function important to management?
6. What is the economic consequence?
7. What is the MD? How is it different from ED?
8. When is the static S/N ratio used?

9. When is the dynamic S/N ratio used?
10. What is the gain? How is it calculated?

References

Breyfogle, F. (2003). *Implementing Six Sigma* (pp. 632–634), Hoboken, NJ: Wiley.

Evans, J. and Lindsay, W. (1999). *The Management and Control of Quality* (pp. 399–405), Cincinnati, OH: South-Western College Publishing.

Fowlkes, W. and Creveling, C. (1995). *Engineering Methods for Robust Product Design* (pp. 34–52), Reading, MA: Addison-Wesley.

Kiemele, M., Schmidt, S.R., and Berdine, R.J. (1999). *Basic Statistics* (pp. 8–6, 8–7), Colorado Springs, CO: Air Academy Press, LLC.

Taguchi, G. and Jugulum, R. (2002). *The Mahalanobis–Taguchi Strategy.* New York: Wiley.

Taguchi, S. (2000). Mahalanobis-Taguchi system, *ASI Taguchi Symposium.* Novi, MI: American Supplier Institute.

17

Design Failure Modes and Effects Analysis

To design is to communicate clearly by whatever means you can control or master.

Milton Glaser

Recognizing, evaluating, and mitigating potential failures is a key aspect of Design for Six Sigma (DFSS). This enables teams to reduce risk within the product, processes, and services that are being designed. Failure modes and effects analysis (FMEA) is a structured method for identifying and mitigating potential failures. Poka yokes are mistake-proofing devices that prevent defects from occurring. Both topics will be discussed in this chapter.

Failure Modes and Effects Analysis

FMEA is a structured approach to assess the magnitude of potential failures and identify the sources of each potential failure. Corrective actions are then identified and implemented to reduce or prevent the potential of a failure occurrence.

FMEA is a prevention-based strategy. It is used to anticipate potential failures, identify potential causes for these failures, prioritize the failures, and subsequently take action to reduce, mitigate, or eliminate these failures. This reduces costs by spending time upfront focusing on preventive action rather than reactive action after the product reaches the customer. Product changes can be implemented more easily and less expensively during the early stages of product development. FMEAs enable the DFSS team to determine what can go wrong, how likely it is to go wrong, and what can be done to prevent it. It is important that a multidisciplinary team conducts the FMEA to ensure a balanced view of the potential failures and their associated causes and impacts.

There are two key types of FMEAs: process and design. Process FMEAs (PFMEAs) are used in Six Sigma to improve processes. Design FMEAs (DFMEAs) are used to improve the design of a product, process, or service by defining and ranking the failure modes for a system. DFMEAs should be conducted for new concepts as a preventive effort to improve the robustness of the concept to failure modes. DFMEAs can also be conducted on each concept generated prior to the Pugh Concept Selection Matrix. A DFMEA should be performed on the concept that emerges as the superior concept from the Pugh Concept Selection Matrix.

In addition, FMEAs can be used to prioritize selected action items from the cause and effect diagram for improvement efforts. The FMEA will identify the causes, assess risks, and determine further steps. A template based on the Automotive Industry Action Group (AIAG) format is provided in Figure 17.1.

The steps to performing an FMEA are the following:

1. Define process steps.
2. Define functions.
3. Define potential failure modes.
4. Define the potential effects of failure.
5. Define the severity of a failure.
6. Define the potential mechanisms of failure.
7. Define current process controls.
8. Define the occurrence of failure.
9. Define current process control detection mechanisms.
10. Define the ease of detecting a failure.
11. Multiply severity, occurrence, and detection to calculate a risk priority number (RPN).
12. Define recommended actions.
13. Assign actions with key target dates to responsible personnel.
14. Revisit the process after actions have been taken to improve it.
15. Recalculate RPNs with the improvements.

The design function requirements represent the form, fit, and functions of the product being designed. These should reflect the requirements that are being fulfilled by the product.

Potential failure modes are the ways in which a product, process, service, subsystem, or component could fail to meet the design intent. Potential failure modes should be listed in technical terms for each function. Examples could include torque fatigue, cracking, or deformation. In addition, since DFMEAs are performed for a system, it is important to recognize that a failure mode in one component can also serve as a failure mode in another component.

The potential effects of failure are the result of the failure mode on the function of the product or process. These describe the effect of the failure modes and should be documented based on how they are perceived by both internal and external customers. Examples of potential effects of failure include degraded performance, injury to the user, noise, and improper appearance.

Potential failure mode and effects analysis
design FMEA

Item ____________
Part description/Number ____________
Core team: ____________

Design responsibility ____________
Key date ____________

FMEA number ____________
Page ____ of ____
Prepared by ____________
FMEA date (orig) ____________

Design function requirements	Potential failure mode	Potential effect(s) of failure	SEV	Class	Potential cause(s)/mechanism(s) of failure	OCC	Current design controls	DET	RPN	Recommended action(s)	Responsibility and target completion date	Action results				
												Actions taken	SEV	OCC	DET	RPN

FIGURE 17.1
Design FMEA.

TABLE 17.1

Severity Criteria

Effect	Criteria: Severity of the Effect	Ranking
Hazardous—without warning	Very high severity ranking when a potential failure mode affects safety and involves noncompliance without warning	10
Hazardous—with warning	Very high severity ranking when a potential failure mode affects safety and involves noncompliance with warning	9
Very high	Process is not operable and has loss of its primary function	8
High	Process is operable, but with a reduced functionality and an unhappy customer	7
Moderate	Process is operable but not easy to manufacture. The customer is uncomfortable	6
Low	Process is operable but uncomfortable with a reduced level of performance. The customer is dissatisfied	5
Very low	The process is not in 100% compliance. Most customers are able to notice the defect	4
Minor	The process is not in 100% compliance. Some customers are able to notice the defect	3
Very minor	The process is not in 100% compliance. Very few customers are able to notice the defect	2
None	No effect	1

The severity of the effect uses a common industry scale from 1, representing no effect, to 10, representing a very severe failure affecting the safe operation of the system. Severity is used by the DFSS team to prioritize the failures. The severity criteria are shown in Table 17.1.

The potential causes/mechanisms of failure represent the design weaknesses that may result in a failure. The cause for each failure mode should be identified and documented using technical terms. The team should be careful not to list these in terms of symptoms. Examples include improper alignment, improper torque applied, and excessive loading.

The occurrence of each cause indicates how likely it is. A common industry scale is used with 1 indicating not likely and 10 indicating inevitable. The occurrence criteria are shown in Table 17.2.

The current design controls are used to identify the mechanisms that are currently in place to prevent the cause of the failure mode from occurring. They also identify the likelihood of detecting the failure before it reaches the customer. The multidisciplinary team should identify testing, analysis, monitoring, or other techniques used to detect failures. Then, each control is assessed to determine how well it is expected to identify or detect the failure mode.

Detection is then determined using a common industry scale by assessing the likelihood that the current design control will detect the cause of the failure mode. The scale is from 1 to 10, with 1 indicating almost complete

TABLE 17.2

Occurrence Criteria

Probability of Failure	Possible Failure Rates	Ranking
Failure is almost inevitable	≥1 in 2	10
	1 in 3	9
High: Repeated failures	1 in 8	8
	1 in 20	7
Moderate: Occasional failures	1 in 80	6
	1 in 400	5
	1 in 2,000	4
Low: Very few failures	1 in 15,000	3
	1 in 150,000	2
Remote: Failure is unlikely	≤1 in 1,150,000	1

TABLE 17.3

Detection Criteria

Detection	Criteria: Likelihood That the Existence of a Defect Will Be Detected by Test Content before Product Advances to Next or Subsequent Process	Ranking
Almost impossible	Test content detects <80% of failures	10
Very remote	Test content must detect 80% of failures	9
Remote	Test content must detect 82.5% of failures	8
Very low	Test content must detect 85% of failures	7
Low	Test content must detect 87.5% of failures	6
Moderate	Test content must detect 90% of failures	5
Moderately high	Test content must detect 92.5% of failures	4
High	Test content must detect 95% of failures	3
Very high	Test content must detect 97.5% of failures	2
Almost certain	Test content must detect 99.5% of failures	1

certainty and 10 indicating almost complete uncertainty. The detection criteria are shown in Table 17.3.

The RPN is the mathematical product of the severity, the occurrence, and the detection. This enables the team to prioritize their efforts on correcting failure modes and their effects. An example FMEA is shown in Figure 17.2.

Once the team have determined the potential failure modes and their associated RPN values, they should recommend actions to address the potential failure modes with high RPN values. These actions could include testing procedures, redesigning the product to eliminate the failure mode, or including backup systems or redundancies. In addition to the recommended actions, the team should assign responsibility and a target completion date

Process function (step)	Potential failure modes (process defects)	Potential failure eflects (KPOVs)	SEV	Class	Potential causes of failure (KPIVs)	OCC	Current process controls	DET	RPN	Recommended actions	Responsible person and target date	Taken actions	SEV	OCC	DET	RPN
Copper strikes	Agnator	100% down potential over reactor	9		Motor, bearing, shaft gearbox	3	None	10	270							
APV	Product Collector Roto Lock	Product not discharging	8		Motor bad, communi cation with scale, jams	4	None	10	320							
APV	Blowers	100% down	8		Plugged filter, bad motor, bad coupling	3	PM on blowers every 4 months	2	48							
Copper nitrate makeup	Discharge valve	Pluggage	8		Powder build up before dissolved	2	Valve design-flush mount	3	48							
Copper nitrate makeup	Mag drive pump	Won't pump	8		Running dry, worn out, motor failure	6	Level indication, recirculation	1	48	Load monitor to be put on, redundant pump						
Copper nitrate makeup	Gate failure	100% down	8		Damage	6	None	1	48	Limit switch, investigate new gate						

FIGURE 17.2
FMEA example.

for each action. This will ensure that responsibility is clear, and it also facilitates tracking. Once the actions have been completed, the severity, occurrence, and detection should be reassessed, and the RPN values should be recalculated. Based on the revised RPN values, the team can decide whether further action is necessary.

What can be seen from the FMEA, which is an important aspect of sustainability, is the RPN reducing after the action items are completed. It is important to understand the severity to a customer. This enables the DFSS team to focus on increasing the capability of the process to in turn improve the process. Reducing the RPN will make the entire process more sustainable by being able to deliver the product at the best capabilities through thorough project management. It is important to maintain the FMEA, so that once a process is improved, this is updated and the team works on the next improvement opportunity identified through the FMEA.

Poka Yokes

Poka yoke is a Japanese term that means "mistake proofing." All inadvertent defects can be prevented from happening or from being repeated.

Poka yokes use two approaches:

- Control systems
- Warning systems

Control systems stop the equipment when a defect or an unexpected event occurs. This prevents the next step in the process from occurring, so that the complete process is not performed. Warning systems signal operators to stop the process or address the issue at the time. The former prevents all defects, and is a more quality control (ZQC) methodology, because an operator could be distracted or not have time to address the problem. Control systems often also use lights or sounds to bring attention to the problem; in this way, again, the feedback loop is minimal.

The methods for using poka yoke systems are

- Contact methods
- Fixed-value methods
- Motion-step methods

Contact methods are simple methods that detect whether or not products are making physical or energy contact with a sensing device. Some of these are commonly known as limit switches, whereby the switches are connected to cylinders and pressed in when the product is in place. If a screw is left out, the product is not released to the next process. Other examples of contact methods are guide pins.

Fixed-value methods are normally associated with a particular number of parts to be attached to a product or a fixed number of repeated operations occurring during a particular process. Fixed-value methods use devices as counting mechanisms. They may also use limit switches or different types of measurement techniques.

Finally, the motion-step method senses whether a motion or a step in the process has occurred in a particular period of time. It also detects sequencing by using tools such as photoelectric switches, timers, or bar-code readers.

Poka yokes are a methodology that can be used as mistake proofing for ZQC to eliminate all defects, not just some. The types of poka yokes do not have to be complex or expensive, just well thought out to prevent human mistakes or accidents.

Poka yokes are also associated with the topic of correct location. This technique places design and production operations in the correct order to satisfy customer demand. The concept is to increase throughput of machines, ensuring that production is performed at the proper time and place. Centralization of areas helps final assemblers; however, to be effective, the most common practice is to develop an effective flow. U-shaped flows normally prevent bottlenecks. Value stream mapping is a key component during this time to establish that all steps occurring are adding value. As a reminder, value-added activities are any activities that the customer is willing to pay for. Another point to remember is not only to have a smart and efficient technique, but also only to produce goods that the customer is demanding, to eliminate excess inventory. Figure 17.3 shows the concept of the importance of mistake proofing.

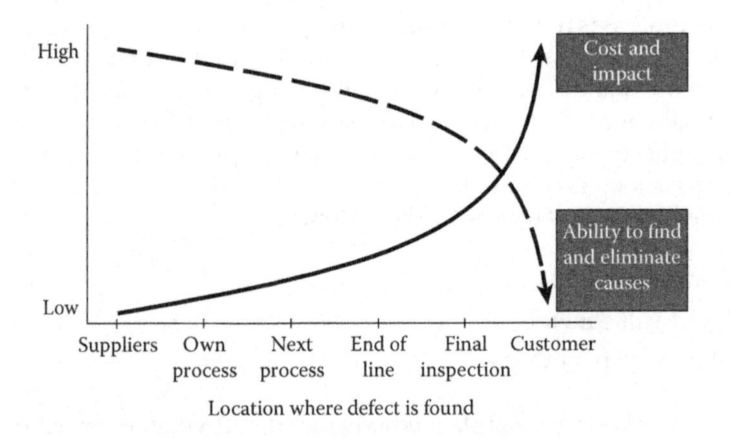

FIGURE 17.3
Impact of defect location.

There are several advantages of mistake proofing first,

- It requires no formal training
- It relieves operators from repetitive tasks
- It promotes creativity and value-added activities
- 100% inspection is internal to the operation
- There is immediate action when problems arise
- It eliminates the need for many external inspection operations
- It results in defect-free work

Errors are the cause

An error occurs when the conditions for successful processing are either incorrect or absent. This results in a defect; however, defect can be prevented if the Defects are the result. Defects are prevented if errors are prevented from happening and the errors are discovered and eliminated.

The distinction between the two is shown in Table 17.4.

How to mistake proof:

- Adopt the right attitude
- Select a process to mistake proof
- Select a defect to eliminate
- Determine the source of the defect
- Identify countermeasures
- Develop multiple solutions
- Implement the best solution
- Document the solution

Which processes should be mistake proofed?

- High error potential
- Complex processes

TABLE 17.4

Mistake-Proofing Examples

Error	Defect
Not setting the timer properly on your toaster	Burnt toast
Placing the "originals" in your copier "face up"	Blank pages
Running out of ink on your date stamp	No date stamp on paper

- Routine "boring" processes
- High failure history
- Critical process characteristic
- High scores in FMEA

How to work on a defect:

- Use problem history
- Use Pareto charts, and so on
- Isolate the specific defect
- Do not be too general
- Do not combine defects
- Make a decision on which defect to work on, based on
    - Severity
    - Frequency
    - Ease of solving
    - Annoyance factor

Figure 17.4 (10 whys for mistake proofing) shows the methodology behind asking "why?" multiple times. Successive checking for positives and negatives is shown in Figure 17.5.

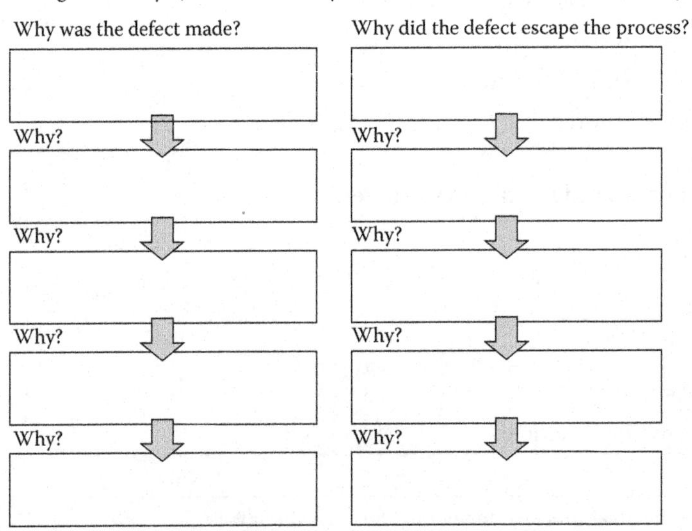

FIGURE 17.4
10 whys for mistake proofing.

Successive checks	Self-check Detection in station	Mistake proof (poka yoke)
Associates check work of previous associate	Associates check own work before passing to the next associate	Automatic check and prevention of defect
Plus: Generally effective in catching defects	Plus: Instant correction possible and more palatable than supervisor check or peer check	Plus: 100% inspection usually with no extra time expense with the benefit of instant correction
Corrective action can only occur after defect is made	Associate may compromise quality or forget to perform self-check	None

FIGURE 17.5
Successive checking.

Conclusion

Mitigating risk is a critical aspect within DFSS. FMEA is a proactive and systematic approach to assessing the magnitude of potential failures, recognizing their sources, and reducing their occurrence.

Questions

1. When should a DFMEA be performed?
2. What is a potential failure mode? Give an example not provided in the chapter.
3. What is the potential effect of failure? Give an example not provided in the chapter.
4. What are the three components of the RPN?

5. A completed FMEA results in the following assessment rating:

   Severity = 7

   Occurrence = 4

   Detection = 3

   What is the RPN for this FMEA?
6. What is a poka yoke?
7. Describe the two approaches for poka yokes.
8. Describe the three methods for poka yoke systems. Give an example of each.
9. Describe the advantages of mistake proofing.
10. What is successive checking? Give an example.

18

Design of Experiments

All life is an experiment. The more experiments you make the better.

Ralph Waldo Emerson

Design of Experiments (DOE)

DOE is a scientific approach to making purposeful changes to inputs (factors) to determine the corresponding changes in the outputs (responses). DOE is an experimental design that shows what is useful, what has a negative effect, and what has no effect. Most of the time, 50% of the factors have no effect.

DOE is used to gain an understanding of the relationship between the input factors and the responses. This relationship can be described through a mathematical model that relates the response to the input factors. The mathematical model can then be used to determine the settings for the input factors to optimize the response.

DOE requires a collection of data measurements, systematic manipulation of variables (also known as factors) arranged in a predetermined way (experimental designs), and control for all other variables. The principle behind DOE is to test everything in a prearranged combination and measure the effects of each of the interactions.

There are several benefits to using DOE. First, it enables teams to determine which factor(s) improve the response variable to an optimal value. In addition, DOE is an inexpensive method that provides equivalent or improved performance compared with other techniques. DOE also helps the Design for Six Sigma team to gain a better understanding of the product or process under consideration, because it defines the magnitude of an effect and identifies how factors react together. Finally, DOE can also be used to systematically demonstrate whether the current product or process is optimal.

The following DOE terms are used:

- Factor: An independent variable that may affect a response, different levels of which are included in the experiment.
- Noise factor: An independent variable that is too expensive or difficult to control.

- Block: A factor used to account for variables that the experimenter wishes to avoid or separate out during analysis. This is used when a group of the experimental runs are more homogeneous than the full set of experimental runs. These are usually selected to allow for special causes, such as a new lot of material or the experiment running over two shifts. The blocks are then included in the experimental order to minimize the impact of the assignable causes, and the team randomizes within the blocks.
- Treatment: Setting or combination of factor levels to be examined during experimentation.
- Levels: Given treatment or setting for an input factor.
- Response variable: The result or output variable from a single run of an experiment at a given setting (or a given combination of settings when more than one factor is involved). A continuous response should be used when possible. The response should be a quality characteristic that addresses the customer's needs and expectations and can be accurately and precisely measured.
- Effect: The relationship between each factor and the response variable.
- Experimental run: A single performance of an experiment given a specific set of treatment combinations.
- n: The number of experimental runs to be conducted in a given experiment.
- Observed value: The response value for a particular test or measurement.
- Experimental error: Variation in the response variable beyond what has been accounted for by the factors, blocks, or other assignable sources in the experiment.
- Replication (Replicate): Repeated run(s) of an experiment at a given setting (or a given combination of settings when more than one factor is involved). This increases the precision of the estimate of the experimental effects; therefore, it is more effective when all sources of experimental error are included.
- Repetition: Occurs when a response variable is measured more than once under similar conditions. This enables the user to quantify the inherent variability in the measurement system.

The purposes of conducting a DOE are to

- Identify Vital Few Xs and how large their main effect is
- Identify interactions between selected Vital Few Xs
- Develop regression model to predict Ys using Xs

TABLE 18.1

Full Factorial DOE

	Factors			Factor Interactions			
Number	A	B	C	AB	AC	BC	ABC
1	−	−	−	+	+	+	−
2	+	−	−	−	−	+	+
3	−	+	−	−	+	−	+
4	+	+	−	+	−	−	−
5	−	−	+	+	−	−	+
6	+	−	+	−	+	−	−
7	−	+	+	−	−	+	−
8	+	+	+	+	+	+	+

- Determine conditions/terms of Xs that optimize Ys
- Decrease variation in the response (*robust design*)

There are two types of DOE: full factorial design and fractional factorial design.

A full factorial DOE determines the effect of the main factors and factor interactions by testing every factorial combination. It analyzes all levels combined with one another, covering all interactions. The basic design of a three-factorial DOE is shown in Table 18.1.

The effects from the full factorial DOE can then be calculated and sorted into main effects and effects generated by interactions.

$$\text{Effect} = \text{Mean value of response when factor setting is at high level}\,(Y_A+)$$
$$-\text{mean value of response when factor setting is at low level}\,(Y_A-)$$

In a full factorial experiment, all of the possible combinations of factors and levels are created and tested.

In a two-level design (in which each factor has two levels) with k factors, there are 2^k possible scenarios or treatments.

- 2 factors each with 2 levels, we have $2^2=4$ treatments.
- 3 factors each with 2 levels, we have $2^3=8$ treatments.
- k factors each with 2 levels, we have 2^k treatments.

Randomization should be used during the DOE to minimize the effect of variation from uncontrolled noise factors. The team should assign treatments to experimental runs such that each run has an equal chance of being assigned a particular treatment. A completely randomized design is an experiment in which each treatment is assigned at random to the full set of experimental units.

The analysis behind the DOE consists of the following steps:

1. Analyze the data.
2. Determine factors and interactions.
3. Remove statistically insignificant effects from the model, such as those with *p* values of less than 0.10, and repeat the process.
4. Analyze residuals to ensure the model is set correctly.
5. Analyze the significant interactions and main effects on graphs while developing the mathematical model.
6. Translate the model into common solutions and make sustainable improvements.

A fractional factorial design quantifies the relationship between influencing factors in a process and any resulting processes while minimizing the number of experiments. A fractional factorial DOE reduces the number of experiments while still ensuring that minimal information is lost. These types of DOEs are used to minimize time and money spent and to eliminate factors that are unimportant.

There are obstacles to using factorial design, such as

- Vague problem statement or unclear objectives
- Inappropriate use of brainstorming
- Ambiguous experimental results
- Money necessary to perform the DOE
- Time spent to conduct the DOE
- Lack of understanding of DOE strategy
- Lack of understanding of DOE tools
- Expectation of immediate results
- Lack of proper guidance/support without validation

The formula for a fractional factorial DOE is

$$2^{k-q}$$

where q is the reduction factor.

The fractional factorial DOE requires the same number of positive and negative signs as a full factorial DOE. Confounding (aliasing) occurs in fractional factorial designs, since not every possible combination is included. Therefore, some factors and interactions are not distinguishable from one another.

The fractional factorial DOE is shown as a matrix in Table 18.2.

TABLE 18.2

Fractional Factorial DOE

												D				
Run	**(I)**	**A**	**B**	**C**	**D**	**AB**	**AC**	**AD**	**BC**	**BD**	**CD**	**ABC**	**ABD**	**ACD**	**BCD**	**ABCD**
1	+	−	−	−	−	+	+	+	+	+	+	−	−	−	−	+
2	+	+	−	−	+	−	−	+	+	−	−	+	−	−	+	+
3	+	−	+	−	+	−	+	−	−	+	−	+	−	+	−	+
4	+	+	+	−	−	+	−	−	−	−	+	−	−	+	+	+
5	+	−	−	+	+	+	−	−	−	−	+	+	+	−	−	+
6	+	+	−	+	−	−	+	−	−	+	−	−	+	−	+	+
7	+	−	+	+	−	−	−	+	+	−	−	−	+	+	−	+
8	+	+	+	+	+	+	+	+	+	+	+	+	+	+	+	+

TABLE 18.3
Three Main Phases of DOE

	Screening	Characterization		Optimization
DOE Type	**Fractional Factorial**	**Fractional Factorial**	**Full Factorial**	**Response Surface Design**
Number of factors	6+	4–10	2–5	2–3
Objective	Identify vital factors	Some interaction	Relationship between factors	Set optimum settings for factors
Estimation	Rough direction of improvement (linear effect)	Main effects and some interactions	All main effects and interactions	Prediction model for output variable (curvature effect)

There are three main phases of DOEs:

1. **Screening phase**: To identify the vital few factors that affect output variable, Y. Modeling is then performed using the vital few.
2. **Characterization phase**: To identify the impact of the vital few, Xs, on output variable, Y.
3. **Optimization phase**: To determine the levels of input variables, Xs, on Y optimization.

The outline for the three main phases is shown in Table 18.3.
It is important to have the following assumptions for DOEs:

- Capable and stable measurement systems analysis (MSA)
- Stable process (statistical process control [SPC])
- Residual analysis (analysis of variance [ANOVA]/regression model)
  - Normality
  - Homoscedasticity (equal variance)
  - Independence

The environmental variables that affect the DOE may result in greater sensitivity of the experiment. The following characteristics are essential for addressing the environmental variables:

- Constant: Variables can be controlled
- Identified but not controlled: Treat as blocking factors

- Not identified or isolated: Randomization of the order of experimental runs (operators, materials, time, and shifts)
- Treated as covariance: Robust design, Taguchi design, analysis of covariance (ANCOVA)

It is important to determine the replication size to make sure the experiment is sufficiently sensitive to the significant effect. If the indicated runs exhaust the available time and resources available, the original plan should be revised, and a reduced model should be used. The number of replicates used should be determined by the power test, which consists of

1. An estimate of the inherent error variation (noise)
2. The size of the smallest variable effect to be significant (signal, resolution)
3. Knowledge of the model to be fitted to the experimental data
4. Required alpha and beta risks
5. Consideration of prior experience with similar processes

The more power a test has, the more sensitive the DOE will be in detecting small differences. As the power increases, the sample size also increases. In addition, as the power increases, the probability of a Type II error, β, decreases.

Conclusion

DOE uses experimental design to develop mathematical models to quantify the relationships between the inputs and the response. It enables DFSS teams to determine the appropriate settings for input factors to optimize the output based on the customers' expectations. The next chapter discusses how to improve the reliability of the design.

Questions

1. What are the benefits of DOE?
2. What is blocking? Give an example of when this could be used that is not mentioned in the chapter.
3. What characteristics should the response factor have in a DOE?
4. Why should the response variable be related to customer needs and expectations?
5. What is it called when an entire experiment is performed more than once?

6. If a factor's effect on the response is dependent on the level of another factor, what is this known as?
7. What is the difference between replication and repetition?
8. In DOE, what is the relationship between a factor and a response variable known as?
9. How is a fractional factorial design different from a full factorial design?
10. Given a two-level design with five factors, how many treatments are there?
11. Why is randomization important in DOE?
12. When is the characterization phase of DOE performed?

19

Reliability Testing

Simplicity is prerequisite for reliability.

Edsger Dijkstra

Reliability is used to quantitatively predict the ability of a product to perform without failure for a specific period of time. It is a method to determine product life before a failure occurs. In Design for Six Sigma (DFSS), reliability is used to quantify the baseline performance of products and components that are being developed. The purpose is to understand the baseline product performance and then improve the product reliability. Reliability can be improved by reducing the variability and sensitivity of the product and components to deterioration and external noise factors.

In product design, it is important for design teams to effectively predict the reliability performance of the overall system as well as the subsystem and components. The reliability data is then used to improve the design and develop appropriate and cost-effective warranty strategies.

Reliability is a measure of the level of certainty that a product will operate without failing for a given period of time. Therefore, reliability is a probability and will always be between 0 and 1.

Product life is a random variable denoted as T. The reliability of the product at time t, then is denoted as $R(t)$ and represents the probability that the product will not fail before time t. The reliability of a product at time t is shown in Equation 19.1.

$$R(t) = P(T > t) \tag{19.1}$$

where:

T = time to failure; continuous random variable

t = specific time

For example, $R(150)$ would represent the reliability of a part after 150 hours. This would be used to determine the probability that the part would operate for longer than 150 hours.

In terms of improving reliability, DFSS teams should focus on the failure modes or the combination of failure modes with high severity. In addition, high risk priority number (RPN) values should be given special attention during the design process, reliability prediction, and reliability testing. There are four key steps for reliability:

1. Determine targets for reliability of the product at the system, subsystem, and component levels.
2. Predict reliability of the product at the system, subsystem, and component levels.
3. Identify gaps between current and target reliability. Based on the gaps, the team should take improvement actions.
4. Predict reliability of the critical components for the product. These are a significant part of the overall system functionality.

In addition, there are four key functions for reliability performance. These are the probability density function (pdf), the cumulative distribution function (cdf), the reliability/survival function, and the hazard function.

The probability density function addresses the number of failures over time, cycles, miles, and so on. This function would fit a histogram of the failure data. The standard normal distribution is the most common example of a probability density function.

The cumulative distribution function is used to determine the probability that the product is not working at any given time.

Conversely, the reliability or survival function is used to determine the probability that the part is still working at any given time.

Finally, the hazard function represents the rate of failure at any given time. The hazard function, $h(t)$, represents the rate at which survivors at time t will fail immediately after time t.

The first phase of the hazard function is considered as the "break-in" period for products. In the second phase, the failure rate is constant for several years. This is known as the "useful life" of the system. Then, as the system gets older, the failure rate increases in the final phase. This is known as the "wearout" period. This failure rate is represented by the bathtub curve, which is shown in Figure 19.1.

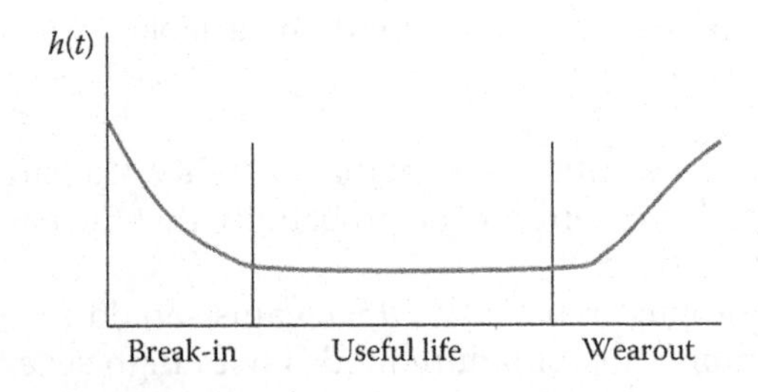

FIGURE 19.1
Phases of a system's lifetime represented in the bathtub curve.

$$\text{Reliability} = n_{operational}(t) / n_{total}(t=0)$$

$$(\text{Number of operational products at time} = t) /$$

$$(\text{Total number of products at time} = 0) \tag{19.2}$$

$$R(t=0) = 1$$

$$\text{Failure probability} = n_{\text{nonoperational}}(t) / n_{\text{total}}(t=0)$$

$$(\text{Number of nonoperational products at time} = t) /$$

$$(\text{Total number of products at time} = 0)$$

$$F(t=0) = 0$$

Failure probability:

$$-F(t) = 1 - R(t) \tag{19.3}$$

Failure density:

$$-f(t) = \frac{\mathrm{d}F}{\mathrm{d}t} \tag{19.4}$$

Failure rate:

$$-\lambda(t) = \frac{f(t)}{R(t)} \tag{19.5}$$

$$= \frac{(\mathrm{d}F / \mathrm{d}t)}{R(t)}$$

$$= \frac{f(t)}{(1 - F(t))}$$

where $f(t)$ is the probability density function (pdf). The average value of this function is the mean time to failure:

- Time 50% failed, only when pdf is symmetrical.
- Time 63% failed, when failure rate is constant.

Area $f(t)\mathrm{d}t$ represents the number of failures in time interval $\mathrm{d}t$. Total sum of all areas $f(t)\mathrm{d}t$ from starting time $t = 0$ to current time t = cumulative distribution function (cdf).

$F(t)$ = failure probability:

By definition, the total area under the pdf = 1; thus, the cdf ranges from 0 (at $t=0$) to 1 (at $t=\infty$).

$$\text{Failure rate} = \frac{\text{(Number of failures in a time interval)}}{\text{(time interval) (number of operational systems at time} = t)} \tag{19.6}$$

Mean time between failures (MTBF) is a well-known metric for repairable systems. Usually, MTBF is only applied when assuming a constant failure rate, because then MTBF = $1/\lambda c$. Note that MTBF $\neq$ lifetime. Weibull distributions can be used to model the failure density $f(t)$:

- When $\beta=1$, the distribution reduces to an exponential distribution: constant failure rate $\lambda=1/\alpha=1/\text{MTBF}$.
- When $\beta=2$, the distribution is called a *Rayleigh probability distribution.*
- When $\beta>3$, the distribution is approximately a normal probability distribution.

Types of Systems

There are three main types of system: series, parallel, and combination. In a series system, all components must be operating for the system to operate. An example of a series system is provided in Figure 19.2. The reliability of n components in a series is given by the product of the respective component reliabilities. This is shown in Equation 19.2.

$$R(t) = R_1(t) \times R_2(t) \times R_3(t) \stackrel{\cdot}{\rightleftharpoons} R_n(t) \tag{19.7}$$

In a parallel system, the system fails if all of the individual components fail. The assumption is that the failure times are independent. An example of a parallel system is given in Figure 19.3. If there are n components in parallel, the formula becomes as shown in Equation 19.3.

$$R(t) = 1 - \left[(1-R_1(t))(1-R_2(t))(1-R_3(t)) \stackrel{\cdot}{\rightleftharpoons} (1-R_n(t))\right] \tag{19.8}$$

C_1 C_2 C_3

FIGURE 19.2
Series system.

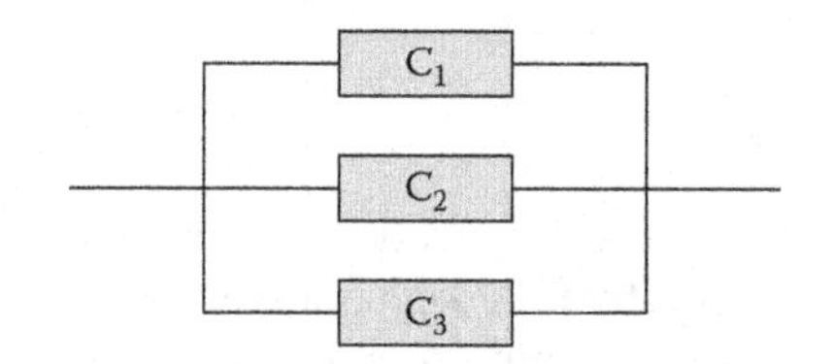

FIGURE 19.3
Parallel system.

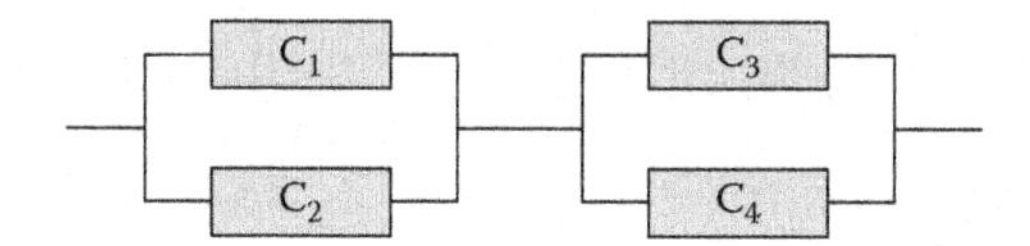

FIGURE 19.4
Combination system.

A combination system includes series and parallel systems in a complex system. An example of a combination system is shown in Figure 19.4.

Example

Suppose the reliabilities of four components are calculated at time $t = 50$ hours. The associate reliabilities are $R_1(50) = 0.90$, $R_2(50) = 0.85$, $R_3(50) = 0.88$, and $R_4(50) = 0.95$. To calculate the reliability of the system, the reliability of the subsystems must be calculated first, as shown in Figure 19.5.

$$R_A(50) = 1 - [(1 - 0.90)(1 - 0.85)] = 0.985$$

$$R_B(50) = 1 - [(1 - 0.88)(1 - 0.95)] = 0.994$$

$$R(50) = R_A(20) \times R_B(20)$$

$$R(50) = (0.985)(0.994)$$

$$R(50) = 0.97909$$

Therefore, the overall reliability of the combination system is 97.909%.

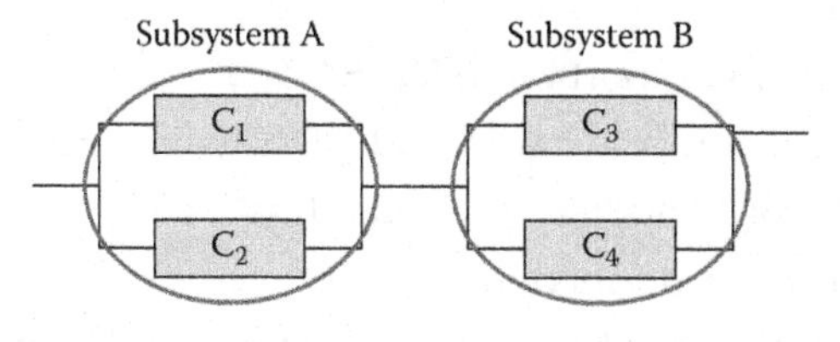

FIGURE 19.5
Combination system example.

Redundant Systems

Backup or redundant systems are used in many situations to improve the reliability of the overall system. A key question in reliability is determining how many redundant systems are necessary to achieve a specific level of reliability. Assuming that components fail independently and operate in parallel, system reliability can be calculated as shown in Equation 19.9.

$$R(t) = 1 - \left[(1 - R_1(t))(1 - R_2(t))(1 - R_3(t)) \rightleftharpoons (1 - R_n(t))\right]$$

$$R(t) = 1 - \left[1 - R_1(t)\right]^n \quad (19.9)$$

Solving for n, we get Equation 19.10.

$$n = \left[\frac{\ln\left[1 - R(t)\right]}{\ln\left[1 - R_1(t)\right]}\right] \quad (19.10)$$

Example

Our customer requires a system that operates with an overall reliability of 0.99 at 200 hours. If we have components that only provide a reliability of 0.88 at 200 hours, how many components do we need in parallel to achieve the required system reliability?

$$n = \left[\frac{\ln\left[1 - R(200)\right]}{\ln\left[1 - R_1(200)\right]}\right] = \left[\frac{\ln\left[1 - 0.99\right]}{\ln\left[1 - 0.88\right]}\right] = \left[\frac{-4.605}{-2.120}\right] = \left[2.172\right] = 3$$

Therefore, three components are needed in parallel to achieve the overall system reliability of 99%. To confirm this, we can calculate the parallel series:

$$R(200) = 1 - \left[1 - R_1(200)\right]^3 = 1 - (0.12)^3 = 1 - 0.001728 = 0.998$$

Questions

1. What is reliability?
2. Why is reliability important in DFSS?
3. To improve reliability, what should DFSS teams focus on?
4. Describe the four key steps for reliability.
5. Describe the hazard function.
6. Give an example of the hazard function for a product you are familiar with.

7. For a parallel system, must all components be operating for the system to operate? Explain your answer.
8. We desire to operate a system with an overall reliability of 0.95 at 300 hours. If we have components that only provide a reliability of 0.85 at 300 hours, how many components do we need to place in parallel to achieve the desired system reliability?

20

Measurement Systems Analysis

The measure of who we are is what we do with what we have.

Vince Lombardi

Measurement systems analysis (MSA) is a specifically designed experiment to determine the variation within a measurement system. Just as a product or process has variation, so does a measurement instrument. MSA provides a systematic approach to evaluating the entire measurement process to ensure the integrity of the data gathered for analysis. In addition, it enables the team then to understand the implications of the error in the measurement system in decision-making regarding the product, process, or service. MSA is performed for the following reasons:

- To ensure the reliability of collected data
- To evaluate the measurement system and to improve the quality of measured data
- To quantify the portion of error in process variation
- To prevent incorrect judgment on defective products from being released to the customer due to measurement error

Gauge repeatability and reproducibility (R&R) is the most popular MSA tool.

An MSA will separate process variation and measurement variation. MSAs are time-consuming and expensive, so they must lead to reduced process variation and improved process control. MSA is an ongoing, periodic, or monitoring tool, which is not meant to "do it once and we are done." Continuous improvement must be ongoing for MSAs to be successful. The impact of any environmental conditions needs to be evaluated or blocked out by randomization.

Figure 20.1 shows the reason why variability plays a large role in processes, and how an MSA is important to reduce the variation.

There are two main components of MSA variation:

1. Variation due to the gauge
    a. Repeatability

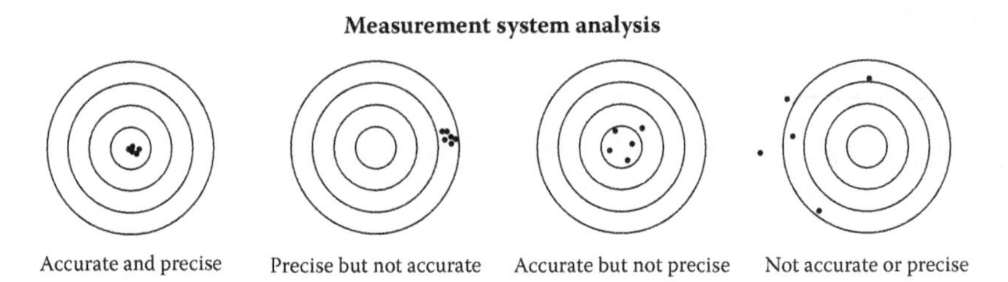

FIGURE 20.1
Precision vs. accuracy.

2. Variation due to operators
   a. Operator
   b. Part by part
   c. Reproducibility

The signal (part–part variation) is the measurement power to distinguish the part–part variation.

The noise is the R&R:

- Repeatability: Variation within same appraiser
- Reproducibility: Variation between appraisers
  - Main effect between operators
  - Interaction effect between operator*part

Either increase the signal and/or reduce the noise to improve the gauge R&R.

There are two main types of data for a gauge R&R:

- Discrete
- Attribute

Discrete gauge R&R plans consist of

- Two or three operators: Experienced operators and blind measurement
- 25–100 parts (larger sample size, higher confidence)
- Two or three repetitions

A balanced gauge R&R must ensure that each operator measures the same part the same number of times.

A gauge R&R attribute agreement analysis has the following components:

- Within operator (repeatability)
- Between operators (reproducibility)
- Each operator versus the standard (individual accuracy)
- All operators versus the standard (overall accuracy)

An attribute gauge R&R sample selection consists of

- Selecting a sample that can represent the process
- Ensuring that the sample size is large enough to detect the meaningful proportion

The guidelines when selecting gauge R&R samples are

- Very difficult to distinguish good/bad 20%–30%
- Difficult to distinguish good/bad 20%–30%
- Somewhat easy to distinguish good/bad 20%–30%
- Very easy to distinguish good/bad 20%–30%

Gauge R&R

Gauge R&R is an MSA technique that uses continuous data based on the principles that

- Data must be under statistical control.
- Variability must be small compared with the product specifications.
- Discrimination should be about one-tenth of product specifications or process variations.
- Possible sources of process variation are revealed by measurement systems.
- R&R are primary contributors to measurement errors.
- The total variation is equal to the real product variation plus the variation due to the measurement system.
- The measurement system variation is equal to the variation due to repeatability plus the variation due to reproducibility.
- Total (observed) variability, σ_{Total}, is an additive of product (actual) variability, $\sigma_{Product}$, and measurement variability, $\sigma_{Measurement}$.

Discrimination is the capability of the measurement instrument to detect small changes (sometimes referred to as *resolution*). It is the number of decimal places that can be measured by the system. Increments of measure should be about one-tenth of the width of a product specification or process variation that provides distinct categories.

Accuracy refers to the clustering of data about a known target.

The *true value* is the theoretically correct value.

Bias is the distance between the average value of the measurement and the true value, the amount by which the measurement instrument is consistently off target, or systematic error. *Instrument accuracy* is the difference between the observed average value of measurements and the true value. Bias can be measured on the basis of instruments or operators. Operator bias occurs when different operators calculate different detectable averages for the same measure. Instrument bias results when different instruments calculate different detectable averages for the same measure.

Precision encompasses total variation in a measurement system, the measure of natural variation of repeated measurements, and R&R.

Repeatability is the inherent variability of a measurement device. It occurs when repeated measurements are made of the same variable under identical conditions (same operators, setups, test units, and environmental conditions) in the short term. Repeatability is estimated by the pooled standard deviation of the distribution of repeated measurements and is always less than the total variation of the system.

Reproducibility is the variation that results when measurements are made under different conditions. The different conditions may be operators, setups, test units, or environmental conditions in the long term. Reproducibility is estimated by the standard deviation of the average of measurements from different measurement conditions.

The *measurement capability index* is also known as the precision-to-tolerance (P/T) ratio (Equation 20.1). The P/T ratio is usually expressed as a percentage and indicates what percentage of the tolerance is taken up by the measurement error. It considers both the repeatability and the reproducibility. The ideal ratio is 8% or less; an acceptable ratio is 30% or less. The 5.15 standard deviation accounts for 99% of measurement system variation and is an industry standard.

$$P/T = \frac{5.15 \times \sigma_{\text{Measurement}}}{\text{Tolerance}} \tag{20.1}$$

The P/T ratio is the most common estimate of measurement system precision. It is useful for determining how well a measurement system can perform with respect to the specifications. The specifications, however, may be inaccurate or need adjustment. Equation 20.2 addresses the percentage of the total variation taken up by the measurement error and includes both repeatability and reproducibility.

$$\%R\&R = \frac{\sigma_{Measurement}}{\sigma_{Total}} \times 100 \tag{20.2}$$

A gauge R&R can also be performed for discrete data, also known as binary data. This data is also known as yes/no or defective/nondefective data. At least 30 data points are still required. The percentages of repeatability, reproducibility, and compliance should be measured. If no repeatability can be shown, there will also be no reproducibility. The matches should be above 90% for the evaluations. A good measurement system will have a 100% match for repeatability, reproducibility, and compliance.

If the result is below 90%, the operational definition must be revisited and redefined. Coaching, teaching, mentoring, and standard operating procedures should be reviewed, and the noise should be eliminated.

Figure 20.2 shows a gauge R&R for a case where there is a decision to be made on which equipment and which employees are sustainable in a factory, based on measurement data. Figure 20.3 provides the goals for % contribution, % study variation, % tolerance, and # of distinct categories. The green regions indicate an acceptable level, yellow indicates opportunities for improvement, and red indicates an unacceptable condition.

The gauge R&R bars are desired to be as small as possible, which causes the part-to-part bars to be larger.

The average of each operator is different, meaning that the reproducibility is suspect. The operator is having difficulty making consistent measurements.

The operator*samples interactions lines should be reasonably parallel to each other. The operators are not consistent with each other.

The measurement by samples graph shows that there is minimal spread for each sample and a small amount of shifting between samples.

The measurement by operators shows that the operators are not consistent, and operator 2 is normally lower than the rest (Figure 20.2).

The sample*operator of 0.706 shows that the interaction was not significant, which is what is wanted from this study.

The percentage part-to-part contribution of 10.81 shows that the parts are the same.

The total gauge R&R percentage study variation of 94.44, percentage contribution of 89.19, tolerance of 143.25, and distinct categories of 1 showed that there was no repeatability or reproducibility, and that this was not a good gauge (Table 20.1). The number of categories being lower than 2 shows that the measurement system is of minimal value, since it will be difficult to distinguish one part from another.

The gauge is poor, based on the fact that the gauge run chart shows that there is no consistency between measurements (Figures 20.2 and 20.4).

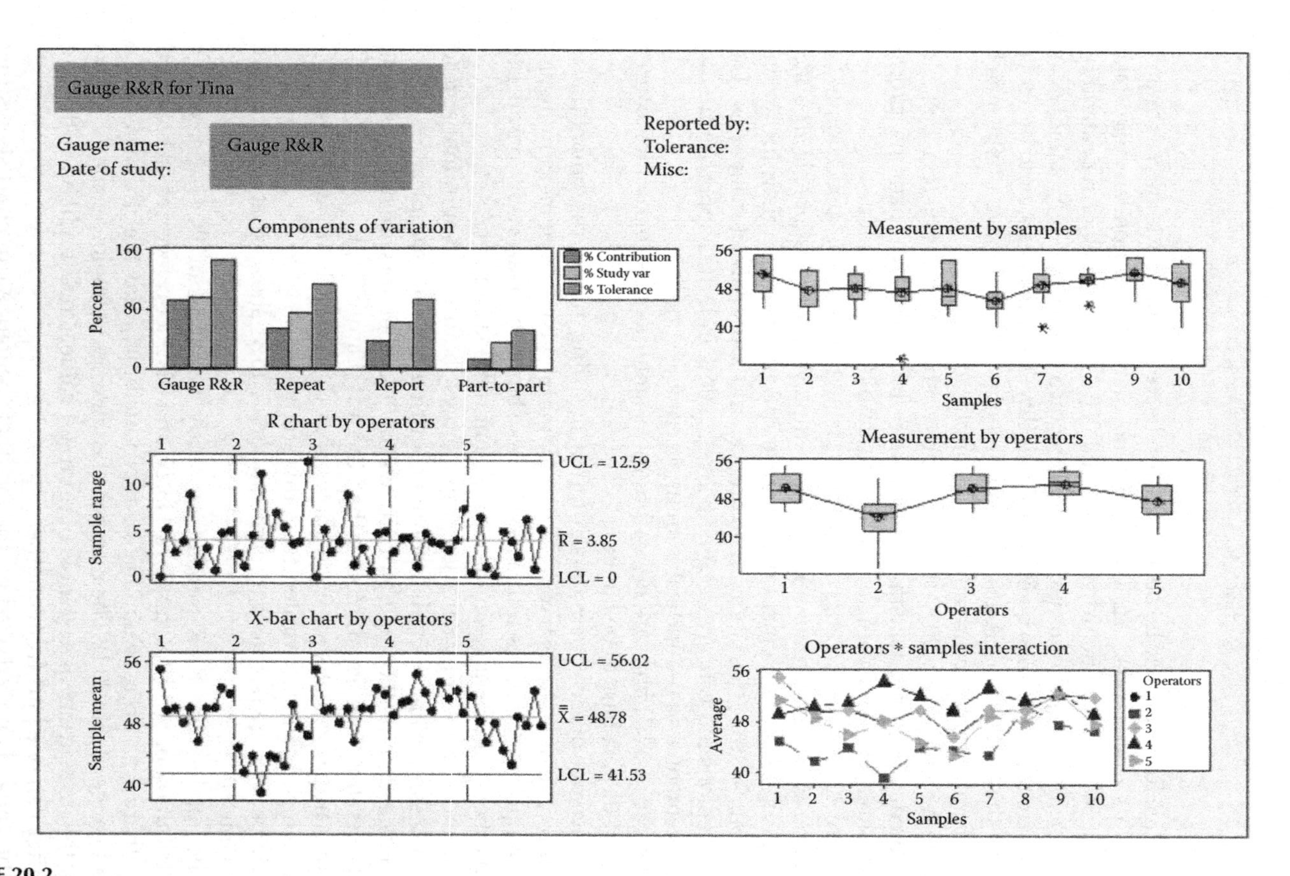

FIGURE 20.2
Gauge R&R example.

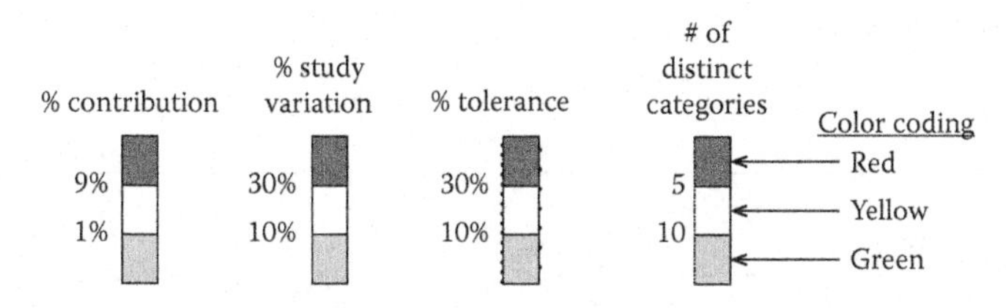

FIGURE 20.3
Gauge R&R results.

TABLE 20.1
Gauge R&R Results

	Process Tolerance = 15			
	Study Var	%Study Var	%Tolerance	
Source	StdDev (SD)	(5.15 * SD)	(%SV)	(SV/Toler)
Total gauge R&R	4.17236	21.4876	94.44	143.25
Repeatability	3.20679	16.5150	72.58	110.10
Reproducibility	2.66929	13.7468	60.42	91.65
Operators	2.66929	13.7468	60.42	91.65
Part-to-part	1.45274	7.4816	32.88	49.88
Total variation	4.41803	22.7529	100.00	151.69
Number of distinct categories = 1				

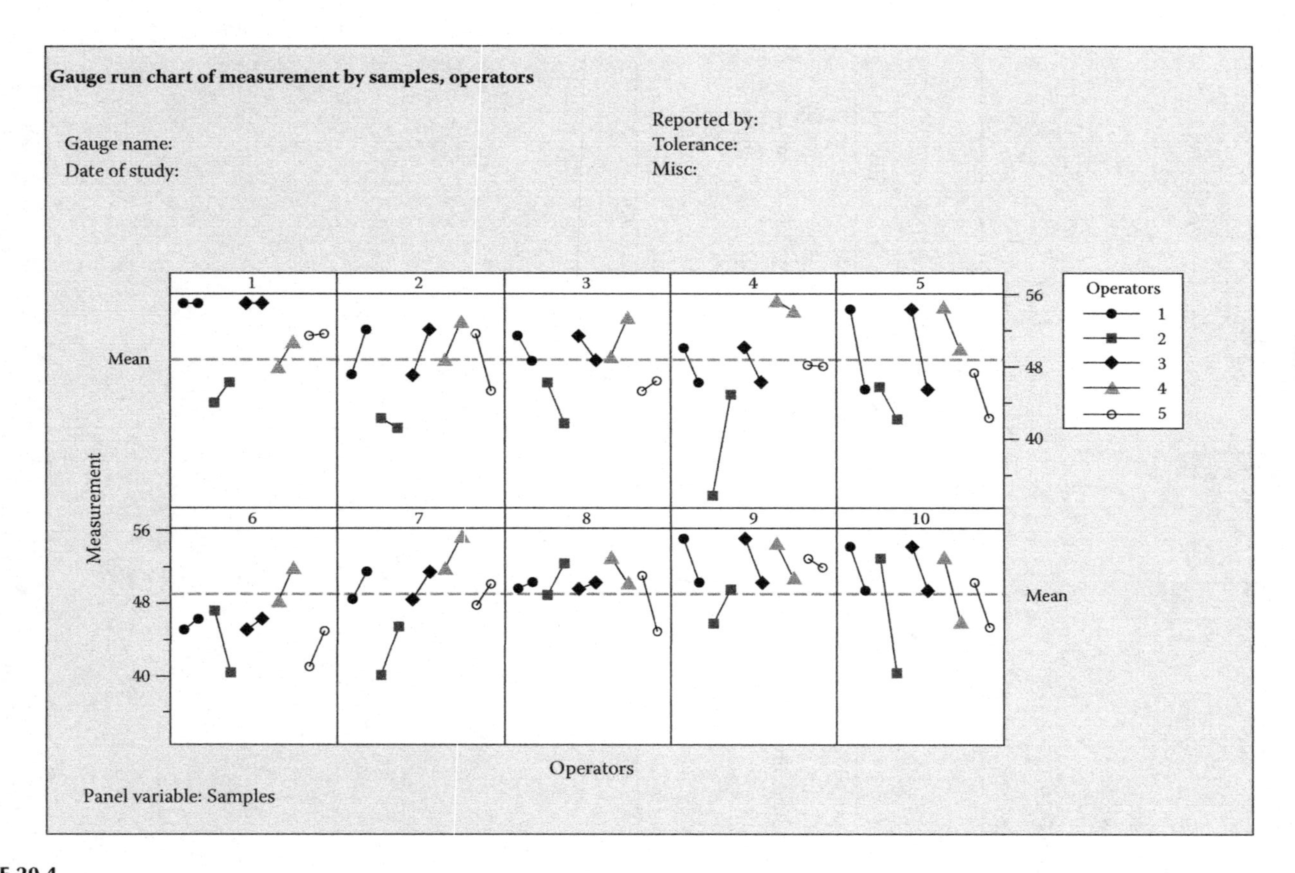

FIGURE 20.4
Gauge R&R run chart.

Conclusions

MSA is an approach to understanding the variation in obtaining measurements. The methodology enables teams to separate process variation and measurement variation. Once an adequate measurement system is in place, the team can be confident in their decision-making based on the measurements. The next chapter discusses capability analysis, which uses collected data to determine whether customer specifications can be met.

Questions

1. What is MSA?
2. What are the two main components of MSA variation?
3. Repeatability is associated with what type of MSA variation?
4. What is between-appraiser variation known as?
5. What is measurement system variation equal to?
6. The following 10 readings were obtained by measuring a 1-in gauge block (a working standard) by a caliper. What is the bias of the caliper in this range?

No.	1	2	3	4	5	6	7	8	9	10
Readings (in)	1.032	1.045	1.035	1.030	1.030	1.035	1.040	1.035	1.040	1.035

7. What are the increments for a measurement system to have appropriate discrimination?
8. How is precision different from accuracy?
9. What is stability?
10. When the percentage R&R is less than 10%, what does this mean?
11. In MSA, what is the term for the ability to produce an average measured value that agrees with the true value or standard being used?
12. What is the term for the variation obtained during an MSA due to differences in people who are taking the measurements?

Technical Design Review: Design Phase

The purpose of the design phase is to determine the new detailed product elements and integrate them to eliminate the problem and meet the customer requirements. The new design is an optimized design to increase robustness and implement error-proof aspects. In addition, tolerance analysis is performed and transfer functions for predictive capability analysis are developed. Statistical process control to review variances and process capabilities is also conducted.

Technical Design Review

Technical design reviews apply a phase/gate approach to technology development. It is a structured and disciplined use of phases and gates to manage product commercialization. The purpose of a technical design review is to prevent design flaws and mistakes in the next phase of development and manage obstacles during a development phase.

A phase is a period of time during which specific tools and best practices are used in a disciplined manner to deliver tangible results that fulfill a predetermined set of technology development requirements. A gate is a stopping point within the flow of the phases of a Design for Six Sigma project.

Two things occur at a gate:

1. A thorough assessment of deliverables from the tools and best practices that were conducted during the previous phase
2. A thorough review of the project management plans for the next phase

A three-tiered system of colors to indicate the readiness of a Design for Six Sigma project to pass through a gate:

1. **Green**: The subsystem has passed the gate criteria with no major problems
2. **Yellow**: The subsystem has passed some of the gate criteria but has numerous moderate to minor problems that must be corrected for complete passage through the gate
3. **Red**: The subsystem has passed some of the gate criteria but has one or more major problems that preclude the design from passing through the gate

Checklists can also be used to list tools and best practices required to fulfill a gate deliverable within a phase. The checklist for each phase should be clear so that it is easy to directly relate one or more tools or best practices to each gate deliverable. Product planning can use the items in the checklist in the development of Program Evaluation Review Technique (PERT) or Gantt charts.

Scorecards are brief summary statements of deliverables from specific applications of tools and best practices. Deliverables should be stated based on the information to be provided including who is specifically responsible for the deliverable and the due date. This helps the technical design review committee manage risk with keeping the project on schedule.

Gate 3 Readiness: Design Phase

- Goal is to take technology that is safe, stable, and tunable into a state of maturity that is insensitive to noise
- Assess the technology's robustness to a general set of noise factors that are likely to be encountered in commercial applications
- Assess the tunability of the critical functional response (CFR) to global optimum targets using critical adjustment parameters

Assessment of Risks

The following risks should be quantified and evaluated against target values for this phase:

- Quality risks
- Delivery risks
- Cost risks
- Capital risks
- Performance risks
- Volume (sales) risks

21

Capability Analysis

I believe more in precision, when you have the capability, like when you see a mosquito fly and you're able to hit it, you're able to hit it with a couple of short sharp shots ... it's a beautiful thing.

Alexis Arguello

Capability Analysis

Industrial process capability analysis is an important aspect of managing industrial projects. The capability of a process is the spread that contains most of the values of the process distribution. It is a way of comparing the "voice of the process" with the "voice of the customer." Process capability provides a measure of how well a process is operating with respect to the customer's expectations.

It is very important to note that capability is defined in terms of distribution. Therefore, capability can only be defined for a process that is stable (has distribution) with common cause variation (inherent variability). It cannot be defined for an out-of-control process (no distribution) with variation arising from specific causes (total variability). It should be determined by the total variation from common causes after special causes have been eliminated. Process capability represents process performance when the process is operating in a state of statistical control. The key need for capability analysis is to ensure that the process is meeting the requirements.

Capability analysis can be calculated with both attribute and variable data. They measure the short- and long-term process capabilities. The key capability indices are C_p and C_{pk}, or P_p and P_{pk} for long-term capabilities. The following formulas are used for capability indices:

$$C_p = \frac{USL - LSL}{6\hat{\sigma}} \tag{21.1}$$

$$C_{pkU} = \frac{USL - \bar{X}}{3\hat{\sigma}} \tag{21.2}$$

$$C_{pkL} = \frac{\bar{X} - LSL}{3\hat{\sigma}}$$

$$C_{pk} = min\left[C_{pkU}, C_{pkL}\right]$$

The value of C_{pk} should be greater than 1.5 for "good" capability. C_p and C_{pk} use a pooled estimate of the standard deviation, whereas P_p and P_{pk} use the long-term estimate of the standard deviation. C_{pk} is what the process is capable of doing if there is no subgroup variability, and P_{pk} is the actual process performance. Normally, C_{pk} is smaller than P_{pk}, since P_{pk} represents both within-subgroup and between-subgroup variability, whereas C_{pk} only represents between-subgroup variability. The main steps to a capability study are

1. Set up the process to the best parameters and identify key process input variables.
2. Identify subgroups.
3. Run the product over a short time span to minimize the impact of special cause variation.
4. Observe the process and take notes throughout.
5. Measure and identify key process output variables.
6. Run capability analyses to review normality, statistical process control, and histograms.
7. Run capability analyses for short-term and total standard deviations.
8. Identify mean shifts and variation.
9. Estimate long-term capability.
10. Develop action plans based on this data.

Setting short-term and long-term goals based on capability analyses will result in successful action plans based on real-time data.
The goal of capability studies is to

- Move the P_{pk} to P_p to center the process
- Move the P_p to C_{pk} to reduce the variation
- Move the C_{pk} to C_p to have random variation

A Six Sigma process has a C_p of 2.00 and a P_{pk} of 1.5.

Control Charts

Control charts may be used to track project performance before deciding what control actions are needed. Control limits are incorporated into the charts to indicate when control actions should be taken. Multiple control limits may be used to determine various levels of control points. Control charts may be developed for costs, scheduling, resource use, performance, and other criteria. Control charts are extensively used in quality control work to identify when a system has gone out of control. The same principle is used to control the quality of work sampling studies. The 3σ limit is normally used in work sampling to set the upper and lower limits of control. First, the value of p is plotted as the center line of a p-chart. The variability of p is then found for the control limits. Two of the most commonly used control charts in industry are X-bar charts and range (R-)charts. The type of chart depends on the kind of data collected: variable data or attribute data. The success of quality improvement depends on (1) the quality of data available and (2) the effectiveness of the techniques used for analyzing the data. The charts generated by both types of data are

- Variable data
  - Control charts for individual data elements (X)
  - Moving range chart (MR-chart)
  - Average chart (X-chart)
  - Range chart (R-chart)
  - Median chart
  - Standard deviation chart (σ-chart)
  - Cumulative sum chart (CUSUM)
  - Exponentially weighted moving average (EWMA)
- Attribute data
  - Proportion or fraction defective chart (p-chart); subgroup sample size can vary
  - Percentage defective chart (100p-chart); subgroup sample size can vary
  - Number defective chart (np-chart); subgroup sample size is constant
  - Number defective (c-chart); subgroup sample size = 1
  - Defective per inspection unit (u-chart); subgroup sample size can vary

The statistical theory useful to generate control limits is the same for all the charts except the EWMA and CUSUM charts.

X-Bar and Range Charts

The R-chart is a time plot useful for monitoring short-term process variations. The X-bar chart monitors longer-term variations, where the likelihood of special causes is greater over time. Both charts use control lines, called *upper* and *lower control limits,* and central lines; both types of lines are calculated from process measurements. They are not specification limits or percentages of the specifications or other arbitrary lines based on experience. They represent what a process is capable of doing when only common cause variation exists, in which case the data will continue to fall in a random fashion within the control limits, and the process is in a state of statistical control. However, if a special cause acts on a process, one or more data points will be outside the control limits, and the process will no longer be in a state of statistical control.

Calculation of Control Limits

- Range (R)

  This is the difference between the highest and lowest observations:

$$R = X_{\text{highest}} - X_{\text{lowest}} \tag{21.3}$$

- Center lines

  Calculate $\bar{X}$ and $\bar{R}$

$$\bar{X} = \frac{\sum X_i}{m} \tag{21.4}$$

$$\bar{R} = \frac{\sum R_i}{m} \tag{21.5}$$

  where:

  $\bar{X}$ = overall process average
  $\bar{R}$ = average range
  m = total number of subgroups
  n = within-subgroup sample size

- Control limits based on R-chart

$$UCL_R = D_4\bar{R} \tag{21.6}$$

$$LCL_R = D_3\bar{R} \tag{21.7}$$

- Estimate of process variation

$$\hat{\sigma} = \frac{\bar{R}}{d_2} \tag{21.8}$$

- Control limits based on $\bar{X}$ -chart

Calculate the upper and lower control limits for the process average:

$$UCL = \bar{X} + A_2\bar{R} \tag{21.9}$$

$$LCL = \bar{X} - A_2\bar{R} \tag{21.10}$$

The values of d_2, A_2, D_3, and D_4 are for different values of n. These constants are used for developing variable control charts.

Plotting Control Charts for *R*- and X-Bar Charts

- Plot the range chart (*R*-chart) first.
- If the *R*-chart is in control, then plot the X-bar chart.
- If the *R*-chart is not in control, identify and eliminate special causes, then delete points that are due to special causes, and recompute the control limits for the *R*-chart. If the process is in control, then plot the X-bar chart.
- Check to see whether the X-bar chart is in control; if not, search for special causes and eliminate them permanently.
- Remember to perform the eight trend tests.

Plotting Control Charts for *MR* and Individual Control Charts

- Plot the *MR*-chart first.
- If the *MR*-chart is in control, then plot the individual chart (*X*).
- If the *MR*-chart is not in control, identify and eliminate special causes, then delete special causes points, and recompute the control limits for the *MR*-chart. If the *MR*-chart is in control, then plot the individual chart.
- Check to see whether the individual chart is in control; if not, search for special causes from out-of-control points.
- Perform the eight trend tests.

Defects per Million Opportunities (DPMO)

Six Sigma provides tools to improve the capabilities of business processes while reducing variation. It leads to defect reduction and improved profits and quality. Six Sigma is a universal scale that compares business processes based on their limits to meet specific quality limits. The system measures DPMOs. The Six Sigma name is based on a limit of 3.4 DPMO.

Figure 21.1 shows a normal distribution, which underlies the statistical assumptions of the Six Sigma model. The Greek letter σ (sigma) marks the

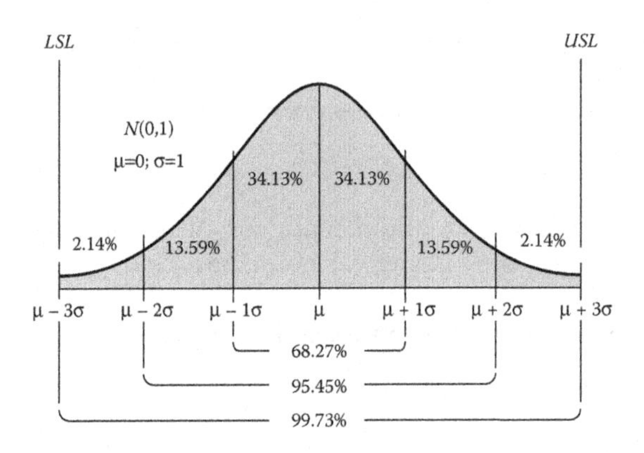

FIGURE 21.1
Areas under normal curve.

distance on the horizontal axis between the mean, μ, and the curve inflection point. The greater this distance, the greater the spread of values encountered. The figure shows a mean of 0 and a standard deviation of 1, that is, $\mu=0$ and $\sigma=1$. The plot also illustrates the areas under the normal curve within different ranges around the mean. The upper and lower specification limits (*USL* and *LSL*) are $\pm3\sigma$ from the mean or within a Six Sigma spread.

Because of the properties of the normal distribution, values lying as far away as $\pm6\sigma$ from the mean are rare, because most data points (99.73%) are within $\pm3\sigma$ of the mean except for processes that are out of control.

Six Sigma allows no more than 3.4 defects per million parts manufactured or 3.4 errors per million activities in a service operation. To appreciate the effect of Six Sigma, consider a process that is 99% perfect (10,000 defects per million parts). Six Sigma requires the process to be 99.99966% within specification limits to produce only 3.4 defects per million, that is, 3.4/1,000,000 = 0.0000034 = 0.00034%. This means that the area under the normal curve within $\pm6\sigma$ is 99.99966% with a defect area of 0.00034%. Six Sigma pushes the limit of perfection. Table 21.1 depicts long-term DPMO values that correspond to short-term sigma levels.

TABLE 21.1
DPMO and Sigma Levels

Sigma Spread	DPMO	Percentage Defective	Percentage Yield	Short-Term C_{pk}	Long-Term C_{pk}
1	691,462.00	69	31	0.33	−0.17
2	308,538.00	31	69	0.67	0.17
3	66,807.00	7	93.3	1	0.5
4	6,210.00	0.62	99.38	1.33	0.83
5	233.00	0.02	99.98	1.67	1.17
6	3.40	0	100	2	1.5

Conclusion

Process capability provides a quantitative measure of process performance for a period of stable operations. Capability compares the "voice of the process" with the "voice of the customer." Process control, however, refers only to the "voice of the process," since it is determined only using data from the process, and customer specifications are not considered. A process is in control if it forms a stable distribution over time. The next chapter discusses statistical process control in detail.

Questions

1. What is necessary for capability to be determined?
2. Calculate C_{pk} given the following information:

   $USL = 12.5$

   $LSL = 12.0$

   Mean = 12.2

   Standard deviation = 0.92
3. What should C_{pk} be for a process to be considered "capable"?
4. What does C_{pk} represent?
5. How is a capability study calculated?
6. What are the goals of capability studies?

22

Statistical Process Control

Facts are stubborn, but statistics are more pliable.

Mark Twain

Statistical process control (SPC) is a methodology that uses graphical and statistical tools to analyze, control, and reduce variation. SPC involves collecting sample data and analyzing this data to detect variation. SPC uses run charts, control charts, histograms, distributions, and confidence intervals for real-time analysis. Using the sample data, teams can establish baselines and dynamically improve process capabilities. Understanding the nature of the variation enables the data to be used for management decisions. SPC enables organizations to effectively take corrective action before waste is produced. Therefore, teams can move from opinion-based to data-driven decision-making. SPC has three primary objectives:

1. Detecting *special cause variation* in the process
2. Measuring the *natural tolerance* of the process (*common cause variation*)
3. Ensuring the process is in control and capable of meeting specifications consistently

SPC provides a means for quantifying and reducing variation, which, in turn, improves product and process design. As the team understands the process behavior, this improves the team's understanding of products and processes. SPC enables real-time monitoring of processes and statistically valid decisions. In addition, by understanding the source of variation, common or special cause, teams can determine when and when not to take action and the appropriate action when needed.

A key aspect of SPC is that it moves away from the traditional approach of detection by producing a product and then inspecting it. SPC is more proactive. It is a preventive strategy that avoids producing waste. The SPC system includes variation, categorizing the sources of variation, control charts, process control, and process capability (Chapter 21).

Process improvement through control charts involves three iterative phases. The first phase is collecting data. Once the data is collected, the second phase is to calculate control limits to determine whether the variation and the mean are stable. If the process is not stable, then special causes will need to be identified and eliminated. The final phase is to assess process capability after the special causes are removed.

Process control and process capability (Chapter 21) are similar, yet different, concepts. The purpose of process control is to enable economically sound decisions. It is important to know when to take action on a system and when not to take action. By taking action when it is not necessary, additional variation from "tweaking" the process is introduced. It is necessary for teams to balance the risks of Type I and Type II errors. A Type I error occurs when action is taken when it is not necessary. A Type II error occurs when the team fails to take action when it should have. These errors occur due to special cause and common cause variation.

Common cause variation is the chance or natural variation in a process. This is the variation that will constantly be active within a system at any time. Therefore, the output of the process forms a distribution that is predictable and stable over time with respect to location, spread, and shape. When there is too much variation in the process, simple statistical techniques can be used to determine the amount of variation; however, detailed and often advanced statistical analyses are needed to determine the causes. To reduce common cause variation, there must be an action on the system by management. Examples of common cause variation include poor machine maintenance, insufficient work instructions, unclear procedures, or normal wear and tear.

Special cause variation is often referred to as *assignable cause*. These causes of variation cannot be explained by a single distribution of the output. Typically, there is a shift in the output due to a specific factor, and the output is unpredictable. These causes can usually be detected by simple statistical techniques. Special cause variation is not common to all operations involved, and discovery is usually by someone directly connected to the operation. Examples of special cause variation include power surges, a poor batch of raw materials, poor machine adjustment, or faulty controls.

A process is in a state of statistical control when only common cause variation exists and special cause variation has been eliminated. When the process is in statistical control, the distribution is stable over time, and the mean and standard deviation do not shift.

There are four main scenarios for process stability and capability:

1. In control and capable: continue the process
2. In control and not capable: improve the process by reducing variation
3. Out of control and not capable: stop the process
4. Out of control and not capable: investigate out of control conditions and/or centering the process

It is important to bring the process back to control before discussing the process capability.

The types of probability distributions are

- Continuous probability distributions
  - Normal distribution: continuous data
  - Variable control charts
- Discrete probability distributions
  - Binomial distribution: represents defective (pass/fail) data
  - Poisson distribution: represents defects/unit data
  - Use attributes control charts

Control Charts

Control charts may be used to track project performance before deciding what control actions are needed. Control limits are incorporated into the charts to indicate when control actions should be taken. Multiple control limits may be used to determine various levels of control points. Control charts may be developed for costs, scheduling, resource use, performance, and other criteria. Control charts are extensively used in quality control work to identify when a system has gone out of control. The same principle is used to control the quality of work sampling studies. The 3σ limit is normally used in work sampling to set the upper and lower limits of control. First, the value of p is plotted as the center line of a p-chart. The variability of p is then found for the control limits. Two of the most commonly used control charts in industry are X-bar charts and range (R-)charts. The type of chart depends on the kind of data collected: variable data or attribute data. The success of quality improvement depends on (1) the quality of data available and (2) the effectiveness of the techniques used for analyzing the data. The charts generated by both types of data are

Variable data:

- Control charts for individual data elements (X)
- Moving range chart (MR-chart); subgroup sample size = 1
- Average chart (X-chart)
- Range chart (R-chart); subgroup sample size is between 2 and 9 ($2 \leq n \leq 9$)
- Median chart
- Standard deviation chart (s-chart); subgroup sample size > 9
- Cumulative sum chart (CUSUM)
- Exponentially weighted moving average (EWMA)

Attribute data:

- Proportion or fraction defective chart (p-chart); subgroup sample size can vary.
- Percentage defective chart (100p-chart); subgroup sample size can vary.
- Number defective chart (np-chart); subgroup sample size is constant.
- Number defective (c-chart); subgroup sample size = 1.
- Defective per inspection unit (u-chart); subgroup sample size can vary.

The statistical theory useful to generate control limits is the same for all the charts except the EWMA and CUSUM charts.

X-Bar and Range Charts

The *R*-chart is a time plot useful for monitoring short-term process variations. The X-bar chart monitors longer-term variations, where the likelihood of special causes is greater over time. Both charts use control lines, called *upper* and *lower control limits,* and central lines; both types of lines are calculated from process measurements. They are not specification limits or percentages of the specifications or other arbitrary lines based on experience. They represent what a process is capable of doing when only common cause variation exists, in which case the data will continue to fall in a random fashion within the control limits, and the process is in a state of statistical control. However, if a special cause acts on a process, one or more data points will be outside the control limits, and the process will no longer be in a state of statistical control.

Calculation of Control Limits

- The range is the difference between the highest and lowest observations:

$$R = X_{\text{highest}} - X_{\text{lowest}} \tag{22.1}$$

- Next, calculate $\bar{X}$ and $\bar{R}$

$$\bar{\bar{X}} = \frac{\sum \bar{X}_i}{m} \tag{22.2}$$

$$\bar{R} = \frac{\sum R_i}{m} \tag{22.3}$$

where:
$\bar{\bar{X}}$ = overall process average
$\bar{R}$ = average range
m = total number of subgroups
n = within-subgroup sample size

- The control for the R-chart are calculated as

$$UCL_R = D_4\bar{R} \tag{22.4}$$

$$CL_R = \bar{R} \tag{22.5}$$

$$LCL_R = D_3\bar{R} \tag{22.6}$$

- The upper and lower control limits for the $\bar{X}$-chart are calculated as:

$$UCL = \bar{\bar{X}} + A_2\bar{R} \tag{22.7}$$

$$CL = \bar{\bar{X}} \tag{22.8}$$

$$LCL = \bar{\bar{X}} - A_2\bar{R} \tag{22.9}$$

The values of d_2, A_2, D_3, and D_4 are for different values of n within subgroup sample size. These constants are used for developing variable control charts, and are provided in Table 22.1.

Plotting Control Charts for Range and Average Charts

- Plot the range chart (R-chart) first.
- If the R-chart is in control, then plot the X-bar chart.
- If the R-chart is not in control, identify and eliminate special causes, then delete points that are due to special causes, and recompute the control limits for the R-chart. If the process is in control, then plot the X-bar chart.
- Check to see whether the X-bar chart is in control; if not, search for special causes and eliminate them permanently.
- Analyze the control chart for out-of-control conditions.

Plotting Control Charts for Moving Range and Individual Control Charts

- Plot the MR-chart first.
- If the MR-chart is in control, then plot the individual chart (X).

TABLE 22.1

Shewhart Constants

n	A_2	A_3	d_2	D_3	D_4	B_3	B_4
2	1.880	2.659	1.128	0.000	3.267	0.000	3.267
3	1.023	1.954	1.693	0.000	2.574	0.000	2.568
4	0.729	1.628	2.059	0.000	2.282	0.000	2.266
5	0.577	1.427	2.326	0.000	2.114	0.000	2.089
6	0.483	1.287	2.534	0.000	2.004	0.030	1.970
7	0.419	1.182	2.704	0.076	1.924	0.118	1.882
8	0.373	1.099	2.847	0.136	1.864	0.185	1.815
9	0.337	1.032	2.970	0.184	1.816	0.239	1.761
10	0.308	0.975	3.078	0.223	1.777	0.284	1.716
11	0.285	0.927	3.173	0.256	1.744	0.321	1.679
12	0.266	0.886	3.258	0.283	1.717	0.354	1.646
13	0.249	0.850	3.336	0.307	1.693	0.382	1.618
14	0.235	0.817	3.407	0.328	1.672	0.406	1.594
15	0.223	0.789	3.472	0.347	1.653	0.428	1.572
16	0.212	0.763	3.532	0.363	1.637	0.448	1.552
17	0.203	0.739	3.588	0.378	1.622	0.466	1.534
18	0.194	0.718	3.640	0.391	1.608	0.482	1.518
19	0.187	0.698	3.689	0.403	1.597	0.497	1.503
20	0.180	0.680	3.735	0.415	1.585	0.510	1.490
21	0.173	0.663	3.778	0.425	1.575	0.523	1.477
22	0.167	0.647	3.819	0.434	1.566	0.534	1.466
23	0.162	0.633	3.858	0.443	1.557	0.545	1.455
24	0.157	0.619	3.895	0.451	1.548	0.555	1.445
25	0.153	0.606	3.931	0.459	1.541	0.565	1.435

- If the *MR*-chart is not in control, identify and eliminate special causes, then delete special causes points, and recompute the control limits for the *MR*-chart. If the *MR*-chart is in control, then plot the individual chart.
- Check to see whether the individual chart is in control; if not, search for special causes from out-of-control points.
- Analyze the control charts for out-of-control conditions.

Figure 22.1 shows a control chart map that can be used to determine the proper control chart to use.

X-Bar and Range Charts

The following components should be used for control chart purposes:

- *UCL*: upper control limit

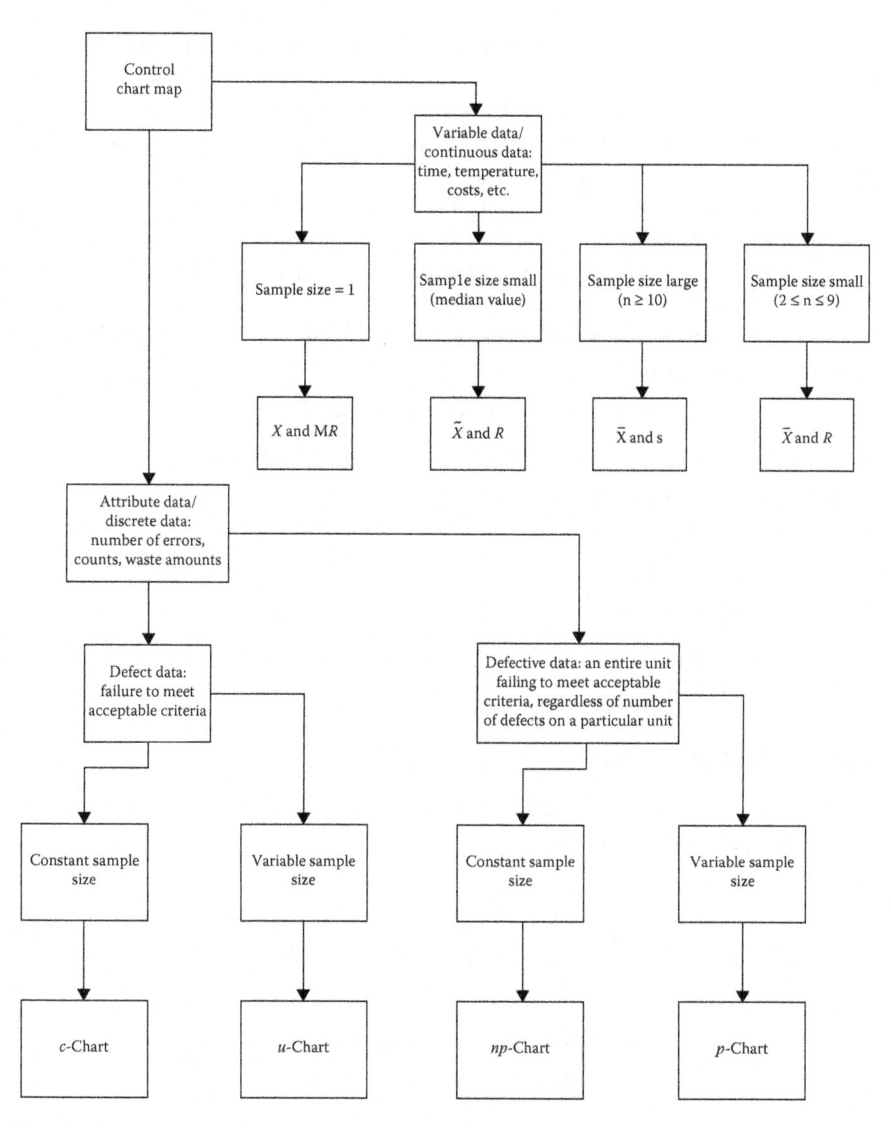

FIGURE 22.1
Control chart map.

- *LCL*: lower control limit
- *CL*: center line—shows where the characteristic average falls
- *USL*: upper specification limit or upper customer requirement
- *LSL*: lower specification limit or lower customer requirement

Control limits describe the stability of the process. Specifically, control limits identify the expected limits of normal, random, or chance variation that is present in the process being monitored. Control limits are set by the process.

Specification limits are those limits that describe the characteristics that the product or process must have to conform to customer requirements or to perform properly in the next operation.

The Shewhart constants listed in Table 22.1 should be used when applicable to the formulas.

There are nine steps to constructing $\bar{X}$ and R-charts:

1. Identify characteristic, measurement method, and sampling pattern.
2. Record data (time).
3. Calculate sample average and sample range R.
4. Calculate grand average and average range $\bar{\bar{X}}$ and $\bar{R}$.
5. If stable, calculate limits.
6. Calculate control limits:
7. Construct control charts.
8. Plot initial data points.
9. Interpret chart with respect to variation: common cause or special cause variation.

Attribute Data Formulas

The following formulas should be used for attribute data when constructing the particular control charts:

p-Chart:

$$p = \frac{\text{Number of defective units}}{\text{Number of units inspected}} \tag{22.10}$$

$$CL = \bar{p} = \frac{\text{Number of defectives in all samples}}{\text{Number of units in all samples}} = \frac{\sum p_j n_j}{\sum n_j} \tag{22.11}$$

$$UCL = \bar{p} + \frac{3\sqrt{\bar{p}(1-\bar{p})}}{\sqrt{\bar{n}}} \tag{22.12}$$

$$LCL = \bar{p} - \frac{3\sqrt{\bar{p}(1-\bar{p})}}{\sqrt{\bar{n}}} \tag{22.13}$$

np-Chart:

$$UCL = n\,\bar{p} + 3\sqrt{\bar{n}\,\bar{p}(1-\bar{p})} \tag{22.14}$$

$$LCL = n\bar{p} - 3\sqrt{n\bar{p}(1-\bar{p})} \tag{22.15}$$

$$CL = n\bar{p} = \frac{\sum x_i}{kn} \tag{22.16}$$

c-Chart:

$$UCL = \bar{c} + 3\sqrt{\bar{c}} \tag{22.17}$$

$$LCL = \bar{c} - 3\sqrt{\bar{c}} \tag{22.18}$$

$$CL = \bar{c} = \frac{\sum c_i}{k} \tag{22.19}$$

where u_j = number of defects in the *j*th sample
u-Chart:

$$CL = \bar{u} = \frac{\text{Number of defects in all samples}}{\text{Number of units in all samples}} = \frac{\sum c_i}{\sum n_j} \tag{22.20}$$

$$UCL = \bar{u} + 3\sqrt{\frac{\bar{u}}{n}} \tag{22.21}$$

$$LCL = \bar{u} - 3\sqrt{\frac{\bar{u}}{n}} \tag{22.22}$$

The rules in Table 22.2 should be used to determine whether a process is out of control for both attribute and variable data.

An example of how to use a control chart to represent whether a process is sustainable after improvements would be to map out the amount of defects per month. A *p*-chart could be useful in this case to see the proportion of defective units per sample produced due to a particular type of error.

TABLE 22.2

Rules for Detecting Out-of-Control Conditions

1. One or more points fall outside of the control limits
2. Two points out of three consecutive points are in the Zone A or beyond
3. Four points out of five points consecutive points in the same Zone B or beyond
4. Nine consecutive points are on the same side of the center line
5. A senses of points are increasing or decreasing
6. Fourteen consecutive points trend upward or downward
7. Fifteen consecutive points are above or below the average

TABLE 22.3

Example Data

Month	Errors	Inspected	Proportion
January	56	100	0.56
February	59	87	0.67816092
March	69	90	0.766666667
April	72	94	0.765957447
May	44	80	0.55
June	39	110	0.354545455
July	46	88	0.522727273
August	25	76	0.328947368
September	20	82	0.243902439
October	15	80	0.1875
November	21	77	0.272727273
December	18	66	0.272727273

Example:

The manufacturing process was unstable, having many defects per month. It was found that the manufacturing equipment was not as reliable as it could be, causing breaks in the system. The defects were monitored per month. A Six Sigma project was identified to improve the reliability of the equipment. The defects were mapped against the number of units inspected to see whether the improvements that were implemented from August through December made an impact (Table 22.3).

Number of Defects Per Month

A *p*-chart was mapped to show the improvements visually (Figure 22.2).

It is clear that there was a mean shift over the last four months that shows a decrease in errors. More points would need to be taken to prove the validity of the improvements over time. The conclusion would be that there was an improvement.

Control limits describe the representative nature of a stable process. Specifically, control limits identify the expected limits of normal, random, or chance variation that is present in the process being monitored. Sustainable processes must follow these rules.

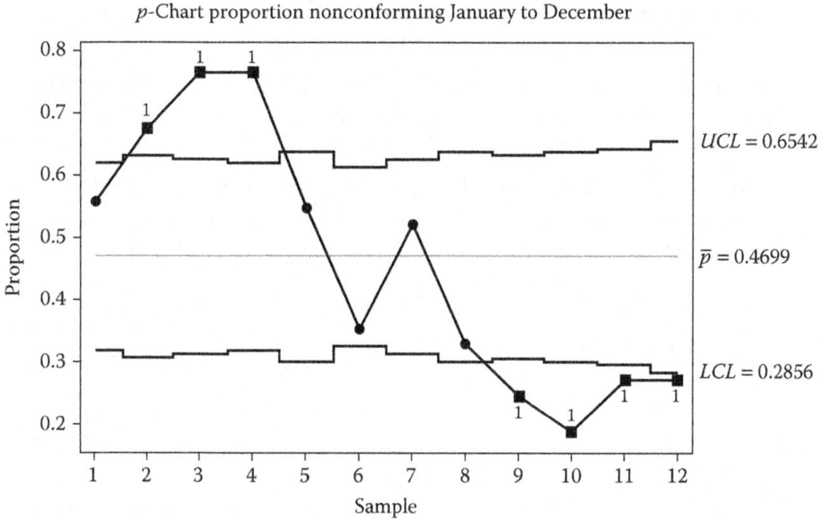

FIGURE 22.2
p-Chart of defects.

Conclusions

SPC is a methodology that enables teams to analyze, control, and reduce variation. SPC enables teams to perform real-time analysis and improve process capabilities. Based on the type of variation, organizations can effectively make management decisions and take corrective action as necessary. In the next chapter, the future of, and challenges to, Design for Six Sigma are covered.

Questions

1. What is SPC?
2. What are the objectives of SPC?
3. What is the difference between process control and process capability?
4. What is a Type I error?
5. What is a Type II error?
6. What is special cause variation? Give an example not included in the chapter.

7. When a process is in a state of statistical control, what can you say about the mean and the standard deviation?
8. The quality of insulation on electrical cable is controlled by checking the number of pinholes in the cable. If the number of cables inspected per day varies, which control chart should be used? Why?
9. Which control chart should be used for the proportion of defective units in a subgroup in which the subgroup size is not fixed and there are multiple opportunities per unit for a defect? Why?

23

Future and Challenges of Design for Six Sigma

The manager accepts the status quo; the leader challenges it.

Warren Bennis

The key difficulties faced when applying Design for Six Sigma (DFSS) are employee engagement and morale while reducing costs and variation simultaneously. The goal for the organization is to maximize the efficiency and effectiveness of processes. The key to doing so is by utilizing employees to make them personally accountable for their improvements.

People want the best for their work life, but they see the improvements being difficult to achieve. "We have always done it this way" is a common theme to resistance.

Managers typically focus on productivity numbers versus improvement activities to show they are meeting customers' demands. It is sometimes forgotten that getting products to the customer is not half as important as getting quality products to the customer.

People begin to ask, "When is there time to work on making a process better when we have to work so hard to get the product out the door?" Many people including managers see quality initiatives taking too much time, which is something they ultimately do not have. Many other employees also feel that they are held responsible for an improvement and do not have the support to get them on the proper track. The entire process is not always well understood and many assumptions are made from people's emotions, opinions, or egos.

An effective way to mitigate these types of assumptions is to use data. Ensuring we have data to prove what is happening is important to keep facts together. It is also very important to understand the motivation for employees to be at work. Employees take pride in the work they do and they want to be satisfied at work. Morale is a key to success.

The DFSS tools and techniques that can help include reducing special cause variation from occurring in the first place while also reducing common cause variation to make procedures more predictable. Reducing the defects from the process will increase the productivity and capacity, which will also make the environment more efficient and safe. To make all of these techniques come together, there are three goals that should be used: (1) maximize profitability, (2) minimize time, (3) minimize costs. In order to achieve these three goals, the following steps should be taken:

1. Understand the customer's problems and how to mitigate those problems
2. Identify the fastest processes with high quality and low cost
3. Reduce any unnecessary or repetitive items
4. Listen to the voice of the customer (VOC) throughout the development process
5. Work on the culture and normal business practices

Engagement and Success Factors

There are several essential elements to designing robust products, processes, and services. There must be a clear, compelling, and urgent reason to change. Urgency should also be a key component to the success of projects. Resistance to change will occur but defining why change is good must be discussed early and often.

Cross-functional leadership is the key to overcoming the challenges of DFSS. Cross-functional leadership must proactively and visibly lead the organization through the change process. This means getting the right leaders to work together to develop a vision of what the organization needs to become and a strategy for getting there. Being proactive while visibly watching change occur will help sustain the Six Sigma journey. Choosing the right leaders for the journey is always key to ensuring sustainability for DFSS. The "right leaders" are those with enough power to lead the change throughout the organization. Leadership must continually communicate and role model the new vision and the strategies. Leadership must break down barriers to making the necessary improvements.

In addition, leadership must engage the people closest to the top priority problems, or the opportunities, to identify, design, develop, plan, and implement the improvements. Leadership must leverage the successes and best practices for making improvements by eliminating waste in other areas. Leadership must help everyone in the organization understand the connection between the improvement activities and the vision of the organization so that the new behaviors become part of the way employees are engaged in continuous improvement and how the business is operated.

Successful DFSS projects require the total immersion of top management. Leaders must create an environment that allows members to participate in the decisions that affect their work, voice honest opinions, and constructively criticize and challenge tradition. This may involve reorienting the organization and changing entrenched behavior. It is critical that the leadership communicates a clear vision, creates a sense of urgency, emphasizes continual training, and stimulates workers and managers alike to engage in the kind of cooperative experimentation that is the cornerstone of a vital, learning organization.

Leadership must identify the vision, align employees at all levels of the organization to that vision, and motivate the employees to achieve that vision. Change requires leadership. Key aspects of providing leadership include clear direction, focused goals and rationale, and willingness to let people make mistakes and express their feelings openly. It is also important to allow employees to make honest mistakes without blame or reprisals and give credit to employees for their ideas, work, and successes.

The elements required for change include change objectives, a vision, and supporting process performance metrics. The infrastructure should also involve a champion, sponsors, facilitators, and teams and appropriate training for each. A concerted effort should be made to develop systems that enable continuous improvement for DFSS to be successful.

As described in Chapter 3, the DFSS methodology requires careful planning with a structured approach. There are several implementation success factors. First, leadership must prepare and motivate people through widespread orientation to continuous improvement, quality, training and recruiting workers with the appropriate skills. A key element of this orientation is to create a common understanding of the need to improve the product development process through DFSS.

Leadership must also structure their DFSS initiative with a focus on employee involvement. This is enabled by leadership pushing decision-making and system development down to the lowest levels. It is also important to share information, manage expectations, and identify and empower champions, particularly engineering managers. Another key aspect is to create an atmosphere of experimentation by tolerating mistakes and being patient. Leadership should install enlightened and realistic performance measures, evaluations, and reward systems and do away with rigid performance goals during implementation. The need to execute pilot projects prior to rolling DFSS implementation out across the organization is essential.

A successful DFSS implementation relies on the organizational framework. Identify champions from top management who will actively support the DFSS projects. Champions need to commit time to attend all design reviews, and be ready to make quick decisions based on their understanding of the benefits that will accrue from implementation.

Technical Design Review: Verify Phase

In the final phase, the new product, process, or service is validated to ensure that customer requirements are met. Statistical confirmation is performed to verify that predictions are met. In addition, control plans, documentation, training, and transition are critical to closing out the project.

Technical Design Review

Technical design reviews apply a phase/gate approach to technology development. It is a structured and disciplined use of phases and gates to manage product commercialization. The purpose of a technical design review is to prevent design flaws and mistakes in the next phase of development and manage obstacles during a development phase.

A phase is a period of time during which specific tools and best practices are used in a disciplined manner to deliver tangible results that fulfill a predetermined set of technology development requirements. A gate is a stopping point within the flow of the phases within a DFSS project.

Two things occur at a gate:

1. A thorough assessment of deliverables from the tools and best practices that were conducted during the previous phase
2. A thorough review of the project management plans for the next phase

A three-tiered system of colors to indicate the readiness of a DFSS project to pass through a gate:

1. **Green**: The DFSS project has passed the gate criteria with no major problems
2. **Yellow**: The DFSS project has passed some of the gate criteria but has numerous moderate to minor problems that must be corrected for complete passage through the gate
3. **Red**: The DFSS project has passed some of the gate criteria but has one or more major problems that preclude the design from passing through the gate

Checklists can also be used to list tools and best practices required to fulfill a gate deliverable within a phase. The checklist for each phase should be clear so that it is easy to directly relate one or more tools or best practices to each gate deliverable. Product planning can utilize the items in the checklist in the development of Program Evaluation and Review Technique (PERT) or Gantt charts.

Scorecards are brief summary statements of deliverables from specific applications of tools and best practices. Deliverables should be stated based on the information to be provided including who is specifically responsible for the deliverable and the due date. This helps the technical design review committee manage risk while keeping the project on schedule.

Gate 4 Readiness: Verify/Validate Phase

Summary analysis and information on

- Patent stance on all new technologies
- Cost estimates for the new technologies
- Reliability performance estimates and forecasts
- Risk summary in terms of functional performance, competitive position, manufacturability, serviceability, and regulatory, safety, ergonomic, and environmental issues

Assessment of Risks

The following risks should be quantified and evaluated against target values for this phase:

- Quality risks
- Delivery risks
- Cost risks
- Capital risks
- Performance risks
- Volume (sales) risks

24

Design for Six Sigma Case Study: Sure Mix Gas Can

Joseph Baumann, Aiswarya Choppali, Sean Flachs, Jeffery Swanson, and Elizabeth Cudney

Design for Six Sigma (DFSS) is a rigorous approach to designing products and services to consistently meet customer expectations. Companies implementing Lean Six Sigma know that many defects are actually created during the design process. DFSS facilitates a redesign of processes—factoring in manufacturing and transactional capabilities from the very beginning—and ensures that end products are "produceable" using existing technology. Additionally, DFSS integrates the engineering and process design functions, enabling concurrent product and process design, and thereby eliminating defects before they can occur.

Project Description

Two-cycle engines are prevalent in the majority of households across the nation. These engines are used in lawn mowers, leaf blowers, weed whackers, chain saws, boat engines, snowmobiles, and a multitude of other tools and devices. Specific to these engines, a precise mixing ratio of gasoline to 2-cycle engine oil, and in some cases marine 2-cycle oil, is required, as specified by the product manual. Without an accurate ratio, consequences can include reduced life span, reduced efficiency, engine failure, lost time, and possibly increased risk to the user due to equipment malfunction. The Sure Mix Gas Can will reduce these issues while also reducing cost.

Project Goals

The project goal for the Sure Mix Gasoline Can is to eliminate the need for additional measuring devices, allowing the customer to purchase 2-cycle oil in any quantity, reducing the hazmat footprint by more than 50%, and increasing mixture accuracy by 25%.

Project Expectations

Customers are currently inhibited by having to mix gasoline with specified amounts of 2-cycle oil or being forced to buy smaller premeasured containers that are more expensive and wasteful. Both existing methods require a great deal of prior preparation or continuous mixing. The purpose of the Sure Mix Gas Can team is to ensure that the future mixing of gas and oil for the purpose of weed whacking, grass mowing, go-karting, and wave running is one of ease and confidence.

Project Boundaries (Scope)

The scope of this project will remain within the United States; more specifically, a trial period would be expected following approval of a lead design product within states of the Midwest region of the United States. This area has been chosen due to the close proximity of supply chains as well as higher than average use of 2-cycle engines over a larger platform of needs: lawn and recreational.

Project Management

The invention and innovation, development, optimization, and verification (I^2DOV) plan was developed with input from each of the group members to ensure that a thorough and complete approach was taken. The product development cycle time and I^2DOV phases were synthesized together.

Identified phases and gates were first established by comparing and contrasting the project outline, the product development cycle time steps, and the I^2DOV phases. From this, the following phases were established: project outline and background, project management outline and plan, voice of the customer (VOC) plan of action, VOC collection, VOC analysis, quality function deployment (QFD), product development, product optimization, and product verification. Each of these phases was then backward planned on the given time line due to the nature of the set schedule and the time constraint given. These dates were agreed on as a group, followed by the assignment of responsible parties for each of the phases.

Agreement on the phases, time line, and responsible parties by the entire group was a vital starting point to ensure that interest was invested by all working on the project. From there, further responsibilities were assigned for deliverables in addition to the phase completion tasks. These deliverables were modeled after the project guidelines, which acted as input from the upper managerial stakeholder of the project. The majority of deliverables at the point of planning were based on the outline, but later focused on specific charts, graphs, or outputs necessary to conduct the Design for Six Sigma (DFSS) project.

From this information, a Gantt or Program Evaluation and Review Technique (PERT) chart was created. Taking software compatibility issues into consideration, it was determined that the PERT chart would be created using Excel. This would not allow automatic calculations and recalculations due to changes in the project, but would allow greater visibility and usability by the team, which was determined to be more valuable. A depiction of the PERT chart used during this project can be found in Figure 24.1.

Lastly, during the creation of the I^{2}DOV roadmap, flexibility was built into the schedule. In this way, tasks were projected to take longer than expected to provide the ability to still meet future deadlines if a previous phase went over schedule. This proved to be a key role in the success of the project, since gathering the VOC through surveys took significantly longer than originally planned. Due to the flexibility built into the schedule, however, the project

Design for Six Sigma Project:
<u>*Sure Mix Gas Can*</u>

	Task	Task #	% Complete	Completion date	% to project
Invent and innovate	*Submit project charter*	*1.1*	100%	31JAN13	5%
	Define goals and mission	1.1.1	100%	24JAN13	
	Define markets and market segments	1.1.2	100%	24JAN13	
	Document technology trends	1.1.3	100%	24JAN13	
	Rough draft project charter	1.1.4	100%	28JAN13	
	Finalize project charter	1.1.5	100%	29JAN13	
	Analyze VOC through results	*1.2*	100%	09APR13	5%
	Contribute 5 survey questions	1.2.2	100%	07FEB13	
	Compile survey questions	1.2.3	100%	12FEB13	
	Finalize survey	1.2.4	100%	12FEB13	
	Administer survey	1.2.5	100%	14FEB13	
	Compile survey results	1.2.6	100%	04APR13	
	Determine VOC population	1.2.1	100%	09APR13	
	Analyze VOC through results	1.2	100%	09APR13	
	Final draft of HOQ	*1.3*	100%	18APR13	2%
	Rank customer needs	1.3.1	100%	11APR13	
	Draft Initiiation of HOQ	1.3.2	100%	11APR13	
	Determine best in class	1.3.3	100%	16APR13	
	Analyze best in class and benchmark	1.3.4	100%	16APR13	
	Compile HOQ with VOC and best in class	1.3.5	100%	16APR13	
	Rough Draft of HOQ	1.3.6	100%	16APR13	
	Final draft of "introduction"	*1.4*	100%	25APR13	7%
	Rough draft of introduction completed	1.4.1	100%	18APR13	
	Review and revision of "introduction"	1.4.2	100%	18APR13	
	Final draft of "project management"	*1.5*	100%	25APR13	8%
	Rought draft of "project management completed"	1.5.1	100%	18APR13	
	Review and revision o f"project management completed"	1.5.2	100%	18APR13	
	Final draft of "invent/innovate" completed	*1.6*	100%	25APR13	8%
	Rought draft of "invent/innovate" completed	1.6.1	100%	18APR13	
	Review and revision of "invent/innovate" completed	1.6.2	100%	18APR13	
	Total % completion for "I2"			100%	

FIGURE 24.1
PERT chart.

for the Sure Mix Gas Can was able to expedite subsequent phases to get back on schedule.

Invent/Innovate

Data drives decisions in product development. Data is needed to define critical and important needs. Face-to-face meetings and a survey were conducted to obtain data. The project's main focus was on customers of lawn mowers, weed whackers, chainsaws, boat engines, and other devices using 2-cycle engines. More specifically, the focus was on a group of customers who perform oil–fuel mixing for a 2-cycle engine. The target customers were decided based on a survey, which indicated that 58% of the people using 2-stroke engines were under 25 years of age, and 86% were male.

Benchmarking

Benchmarking was used to compare the features of different gas cans. Gas cans from three different manufacturers were considered, and their pros and cons were assessed, as shown in Table 24.1.

Although our competitors had many positive aspects, they did not satisfy the criterion of elimination of mixing and measuring.

Voice of the Customer

To understand the specific needs of customers, a survey was conducted. From the 96 relevant responses, the team was able to infer that 60% of the customers measured oil for 2-stroke engines rather than buying it. The survey concentrated on specific improvements for appropriate measurements of

TABLE 24.1

Benchmarking Analysis

No.	Competitor 1	Competitor 2	Competitor 3
1.	Push button control	Wide base for safety	Made of durable plastic
2.	Flow stops automatically	Galvanized steel	Spill proof
3.	Large neck opening	Durability	Not costly
4.	View stripes front and back		

the oil–fuel mix for a 2-cycle engine. The customer requirements were classified under critical and important needs.

- Critical needs
  - Elimination of oil and fuel measurements
  - Elimination of mixing of oil and fuel
  - System for multiple oil–fuel ratios
  - Visual monitoring of fuel level
- Important needs
  - Safety of personnel handling equipment
  - Easy transportation
  - Durable material
  - Easy to handle and clean
  - Avoidance of spills

Affinity Diagram

An affinity diagram was used to classify all the customer needs and requirements into groups to help identify appropriate technological system requirements. Customer requirements that require similar technological requirements are grouped, as shown in Table 24.2, and assessed. Functional requirement is then decided based on the group of customer requirements.

House of Quality

The customer requirements were studied and sorted using an affinity diagram. Using new, unique, and difficult (NUD) needs, the house of quality (HOQ) was developed. NUD requirements were studied in descending order from the highest-rated customer requirement to the lowest-rated customer requirement. The functional requirements were decided based on customer requirements. The HOQ is shown in Figure 24.2.

TABLE 24.2

Classification of Customer Requirements

Oil–Fuel Mix	Measurement	Clean
Dropping premixed packets Specific mix packets depending on fuel quantity Ratios for large quantities	Completely premixed fuel and oil. No mixing, no measuring, no mess Some read out of the mixture ratio	Some way of making sure oil does not get anywhere and is easily cleanable No drips, no spills Completely premixed fuel and oil. No mixing, no measuring, no mess

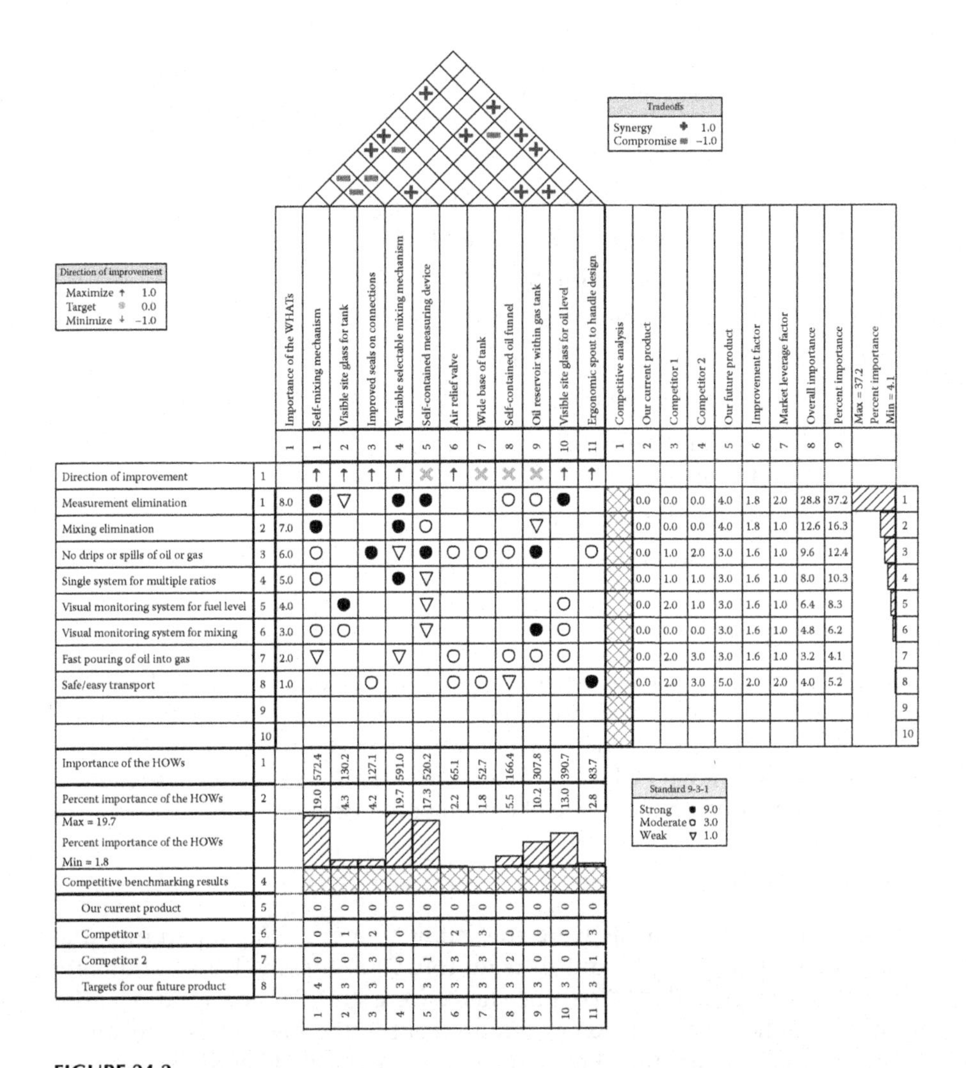

FIGURE 24.2
The house of quality.

Depending on the weight charts of functional requirement, the top three features that the gas can must incorporate are (Figure 24.3)

- Variable selectable mixing mechanism: 18% by weight
- Self-mixing mechanism: 17% by weight
- Self-contained measuring device: 16% by weight
- Oil reservoir within the gas tank: 13% by weight

Further, these functional requirements were incorporated into the design to develop the technology.

17%	Self-mixing mechanism
6%	Visible site glass for tank
7%	Improved seals for connections
18%	Variable selectable mixing mechanism
16%	Self-contained measuring device
4%	Air relief valve
3%	Wide base of tank
6%	Self-contained oil funnel
13%	Oil reservoir within gas tank
5%	Visible site glass for oil level
4%	Ergonomic spout handle design

FIGURE 24.3
Functional requirements.

Kano Analysis

The Kano Model defines three types of quality requirements and their effects on customer satisfaction. The different customer requirements are identified and classified under one-dimensional, expected, and exciting quality (Figure 24.4).

Among the customer requirements, the elimination of measurement and mixing was given the highest priority (Figure 24.5). As well as the necessary product characteristics, these two requirements are to be incorporated based on customer review.

Design for X Methods

The design for X concept was used next in the development of the Sure Mix Gas Can for the NUD requirements. The NUD requirements were addressed

One-dimensional quality	Expected quality	Exciting quality
Self-mixing mechanism	Durable material	Ergonomic spout to handle design
Visible site glass for tank	Good air seal	Visible site glass for oil level

FIGURE 24.4
Kano analysis.

Customer importance	Maximum relationship	Customer requirements (explicit and implicit)
8	9	Measurement elimination
7	9	Mixing elimination
6	9	No drips or spills of oil or gas
5	9	Single system for multiple ratios
4	9	Visual monitoring system for fuel level
3	9	Visual monitoring system for mixing
2	3	Fast pouring of oil into gas
1	3	Safe/easy to transport

FIGURE 24.5
Weighted customer requirements.

at the system level of the Sure Mix Gas Can. The design for X parameters that were addressed in the concept and design process were produceability, assembly, reliability, testability, and friendliness to the environment.

With regard to produceability, the Sure Mix Gas Can does not incorporate any radical design changes from existing plastic molding technology. The gas can itself is similar to current plastic gas cans on the market. In fact, removing the spout from the gas can would make it almost identical to any other commercially sold gas can. The spout for the Sure Mix Gas Can required the most thought regarding produceability. First, the cap requires slightly more exact molding due to the different plastic keys designed to prevent the user from putting the cap on incorrectly. Second, the spout has a flow gate on the oil flow side that needs to be molded or machined to accept the measured gate valve.

Assembly of the gas can is fairly straightforward, with a total of eight pieces. The caps, spout, and can are almost foolproof. The item that could cause most issues during assembly is the measured flow gate valve, as it requires inserting a valve into the plastic cap without destroying the oil–gas separating partition in the cap.

Reliability is almost ensured for the Sure Mix Gas Can. Since the goal of the design was to create a gas can that could provide a multitude of fuel mixes for 2-stroke engines, the metered valve, when placed at the appropriate indicator, will provide a flow of oil into the gasoline stream that will mix itself as it goes through the spout and into a gas tank. The only way not to get the appropriate mixes would be not to have enough gas or oil in the gas can. The method for mitigating this problem was to design the gas can out of translucent plastic or to include translucent plastic strips on the gas can itself.

Testability is fairly direct for the Sure Mix Gas Can, as either the can produces the required mix or it does not. The design can be tested in two ways. The first way is to test whether the can produces the correct mix ratio of oil and gas. The second method is to test the flow of oil from the container when the valve is set to a certain level; in that way, the oil could be reused, and it would save money in the testing phase.

Finally, to design the gas can to be environmentally friendly, the team had to ensure that the can would not leak or break when subjected to severe shocks and would also be recyclable. In the design phase, the team chose to use plastic, so that the container could be easily molded and would be impervious to oil or gas leaks. To ensure that there would be no leaks, rubber gaskets would be included on the spout cap and oil cap. However, the team recognized that the gas can would lose structural integrity if exposed to ultraviolet light for six months or longer. Also, since the container is made from high-density polyethylene, the gas can would be recyclable.

Concept Generation

Concept generation for the Sure Mix Gas Can took three forms: benchmarking, brainstorming within the group, and patent studies. To begin concept generation, patent studies and benchmarking were first used. There was a patent issued in 1959 regarding an oil–gas mixing funnel for 2-stroke engines (patent number 2,902,062). The group considered this method of mixing. Benchmarking was the second approach, with several different gas cans evaluated on their merits to work with 2-cycle engine oil mixing. No gas cans addressed 2-stroke oil mixing. The next step was brainstorming within the group, which yielded the concepts of having a mixing funnel integrated into the gas can and two separate cans mixing in a tube. Finally, the team created seven concepts, which had integrated funnels, mixing tubes, and integrated mixing spouts.

Criteria	Concept A	Concept B	Concept C	Concept D	Concept E	Concept F	Concept G	Datum
Measurement elimination	+	S	+	+	+	+	–	
Mixing elimination	+	–	+	–	–	+	+	
No drips or spills	S	–	S	–	–	–	S	
Single system for multiple ratios	+	–	+	–	–	S	S	
Visual monitoring system for fuel level	+	+	–	–	+	–	–	
Confidence in mixing	–	+	–	–	+	–	–	
Fast pouring of oil into gas	+	–	+	+	–	+	–	
Safe/easy to transport	S	S	S	S	S	S	S	
Cost	–	–	–	–	–	–	–	
Sum of (+)	5	2	4	2	3	3	1	
Sum of (–)	2	5	3	6	5	4	4	
Sum of (S)	2	2	2	1	1	2	3	

FIGURE 24.6
Pugh's Concept Selection Matrix.

After the team developed the various concepts, Pugh's Concept Selection Matrix (Figure 24.6) was used to compare the designs with the benchmark, a 1-gallon highlyrated gas can. As shown by the matrix, the two designs that came out superior were Concepts A and C. However, the matrix also shows that all of our concepts cost more than the datum. The team reconciled this cost difference, as our concepts have more components, and the goal was also to eliminate multiple gas cans for different mix ratios.

Concept A (Figure 24.7) focused on mixing the 2-stroke oil in the spout with a Venturi system to allow the correct flow of oil to make the desired oil–gas ratio. The can is 2.5 gallons in volume, with a clear sight glass to enable seeing how much gas and oil are in the can, and a separated oil compartment. This design had the highest selection score but failed on cost. The design was also more difficult in terms of manufacturability. Aspects from Concept A were integrated into the final design.

Concept B (Figure 24.8) mixed the oil and gas together by means of a collapsible measuring cup. The idea was to have a collapsible cup that would

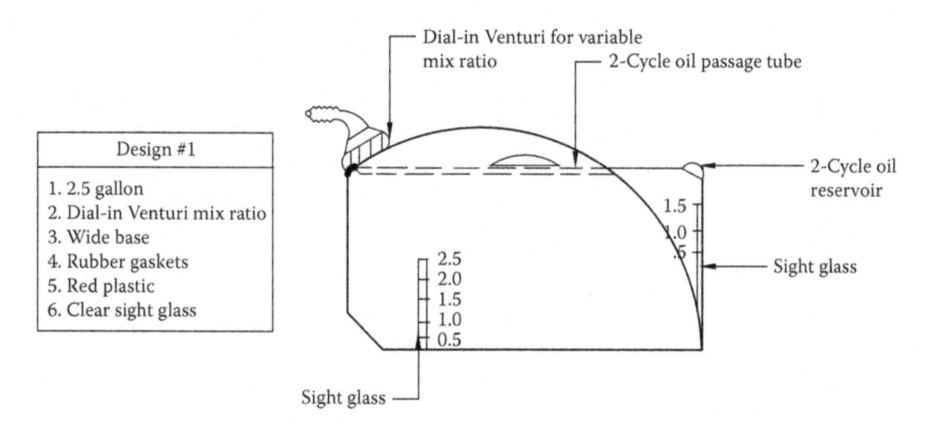

FIGURE 24.7
Concept Design A.

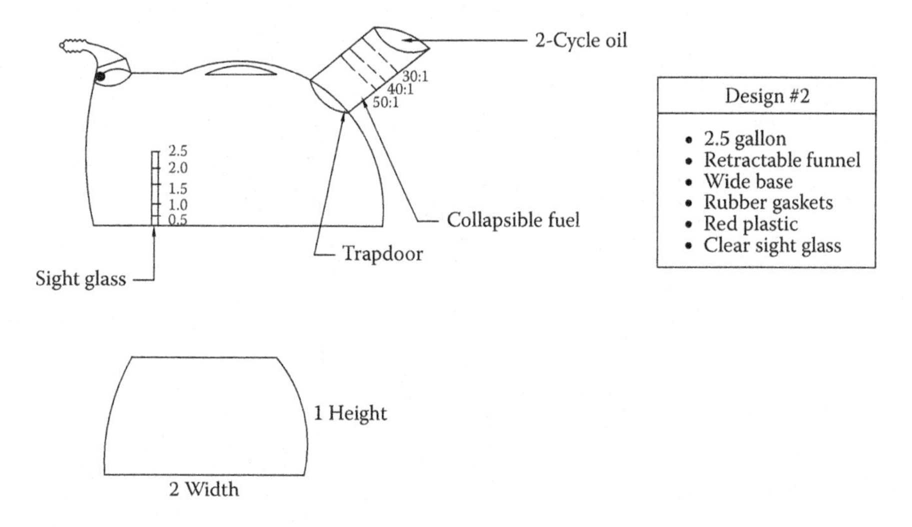

FIGURE 24.8
Concept Design B.

measure the appropriate ratio for the gas can capacity and then open a trapdoor to allow the gas and oil to mix. This design was rejected, as it failed to allow a variable mix ratio for one gas can.

Concept C (Figure 24.9) also focused on integrating the mixing of the oil and gas in the spout of the gas can. The gas can contained a separate compartment for oil and gasoline. The spout had a thumb lever that would allow the user to press the lever down to get the appropriate amount of oil flowing into the spout to achieve the correct oil–gas ratio in the spout. This design was rejected because there was no way to see how much oil or gas was left

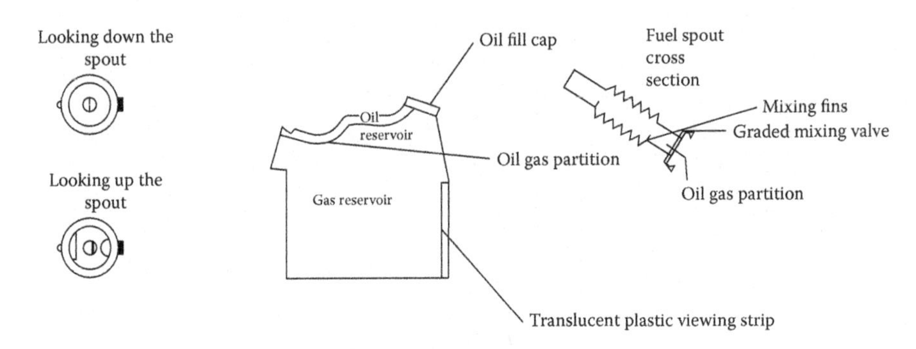

FIGURE 24.9
Concept Design C.

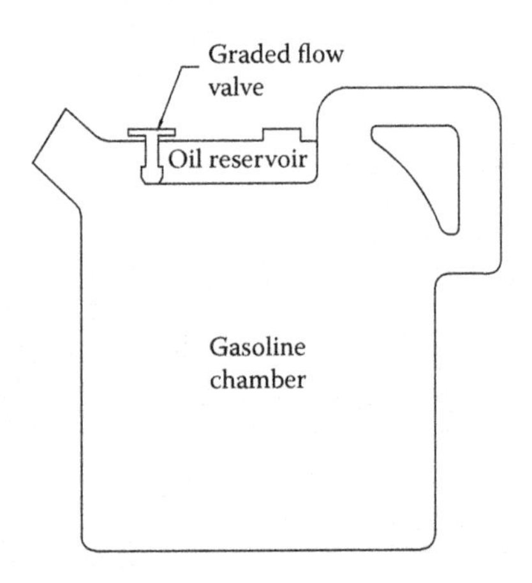

FIGURE 24.10
Concept Design D.

in the gas can, potentially causing inappropriate mixes if there was gasoline flowing but no oil. This spout concept was considered more easily manufactured than Concept A but lacked Concept A's reliability.

Concept D (Figure 24.10) had a separated oil compartment with a graded flow valve. The idea behind Concept D was that the user would turn the valve to get the appropriate oil–gas mix; however, oil could mix with the gasoline after pouring was complete or not mix at all if there was not enough oil in the chamber.

Concept E (Figure 24.11) had an integrated measuring funnel to allow the user to pour the appropriate amount of oil required to mix with the gasoline for the desired ratio. Additionally, the design was made for usability and reducing spills due to the lower stopcock, so that the user would not have to bend as far to pour the oil–gas mix.

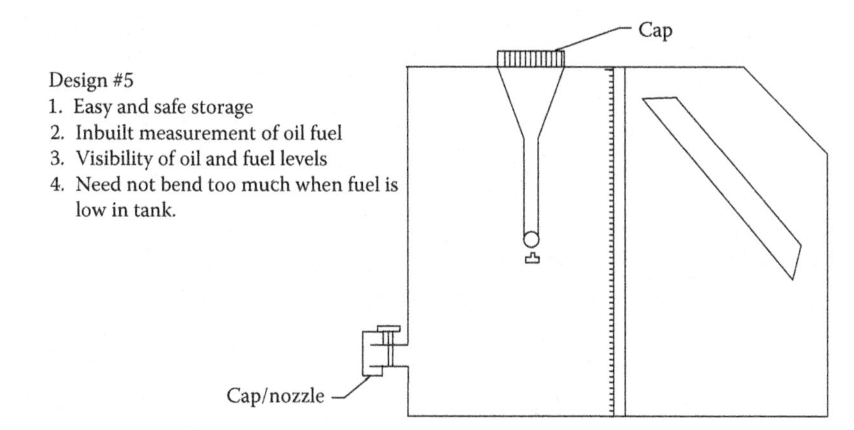

FIGURE 24.11
Concept Design E.

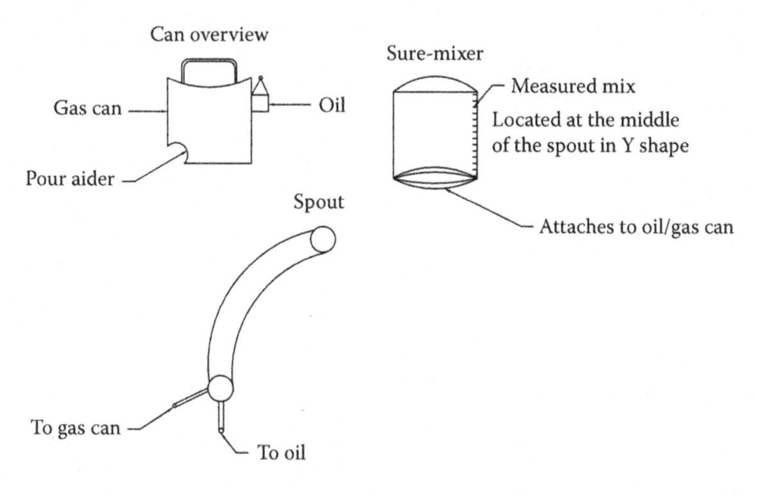

FIGURE 24.12
Concept Design F.

Concept F (Figure 24.12) mixed the oil and gas by means of the spout. The oil would be measured out in a separate external container for the appropriate mixture ratio for the gasoline container. However, there was no way to control the flow of the oil should only a partial pour be required.

Concept G (Figure 24.13) achieved the oil–gas mix ratio with a gravity feed. The oil and gas had separated reservoirs that fed into flow-regulated tubes, which then caused the oil and gas to mix in the spout or the receiving gasoline tank. Though the gravity feed system had merits, it significantly increased the possibility of spills, as the Sure Mix container would have to be lifted to a higher level than the tank that needed to be refilled.

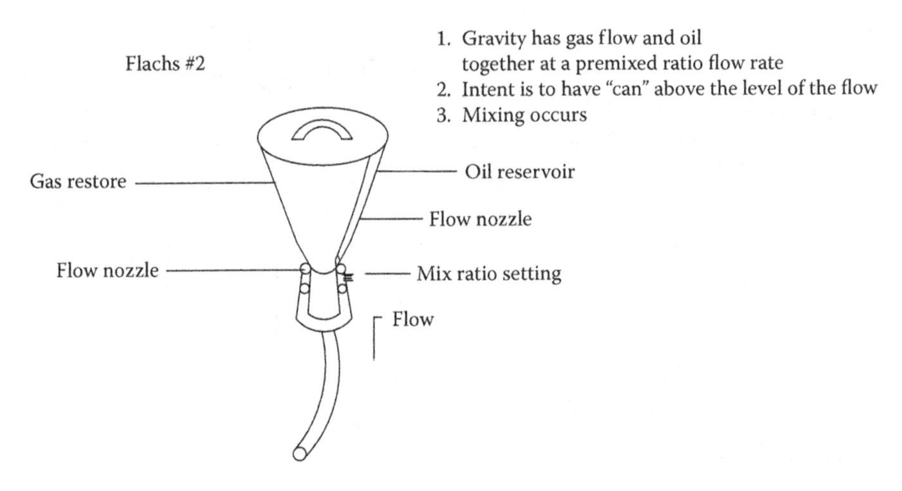

FIGURE 24.13
Concept Design G.

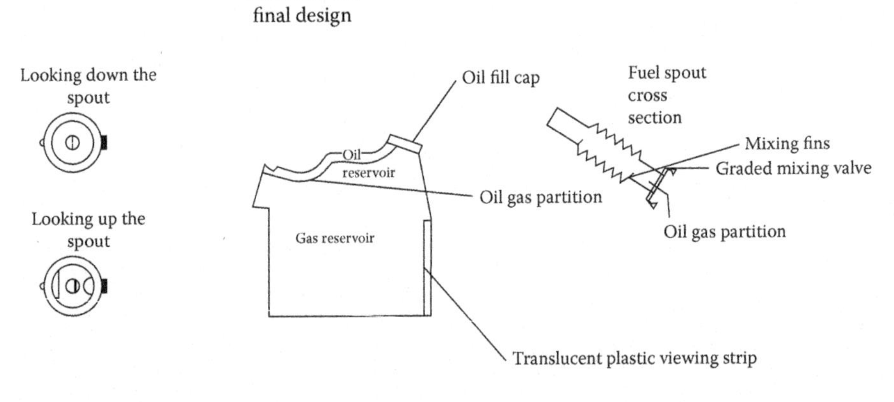

FIGURE 24.14
Final design.

Concept H (Figure 24.14) was the last conceptual design, which incorporated elements from Concept A and Concept C. It was decided to have the container made of translucent plastic to allow the user to see how much oil and gas were in the container at any time. Also, the oil compartment was integrated into the handle, as that section is not usually filled with gasoline, and this would result in the oil being gravity fed if the measuring gate was open. The mixing occurred in the pour spout, and the user could adjust the rate at which the oil came out with a graded valve. This design allows the user to have the appropriate oil–gas ratio during a pour, with one container capable of providing different ratios of oil–gas mixes.

Technology Modeling

Technology modeling was limited, since the team possessed neither the equipment nor the means to model the technology. However, the team did brainstorm several ideas on noise and variance regarding the final design. The system variance will come from the subsystem for regulating the flow of oil into the pouring gasoline stream. One of the noise factors is the oil viscosity. To reduce the noise from oil viscosity, the team recommends a usable operating temperature range so that the graded mixing valve will read correctly. Another factor in noise suppression is the integration of the oil reservoir into the gasoline can. This will prevent extraneous parts from wearing, getting misplaced, or accidentally breaking.

As such, the final concept is fairly robust, as it can offer the consumer one oil–gas mixing can. Also, should the user not be satisfied with the mixes, the user can shut off the oil flow valve and still mix and use the gasoline can.

Robustness/Tunability

With the customer population covering all parts of the United States, there will be many different noise factors that may affect the newly developed gas can. Because it will not be possible to remove these outside factors, the robustness of the system must be addressed. To this end, the team first identified the critical functional responses (CFRs). For the purposes of this project, the top response, the variable selectable measuring mechanism, was selected. The metric that would be used to measure this would be the accuracy, in ratio parts of gas to oil, of the resulting fuel coming out of the spout. This would be compared with a finite goal, and the acceptable range would be identified through the resulting engine performance ranges.

One example of a noise factor that could affect this CFR is the temperature at which the process is taking place. This would be considered an external noise, in that it comes from an external source. An increase in temperature will reduce the viscosity of the oil, making it less thick and allowing it to flow through the restrictors more easily. If the temperature is low, then the opposite will occur, in that not enough oil will mix with the gas, creating an inaccurate fuel to oil ratio. Other noises that would affect the gas can include wearing down of the clicking mechanism for maintaining the desired gas/oil mixture setting (deterioration), the viscosity of the brand of 2-cycle oil or the type of oil used (external), and the amount by which the lever must be opened to release the flow of gas, which would affect the Venturi pull effect of the oil (unit-to-unit noise).

These noises were captured in a p-diagram, as shown in Figure 24.15. Designed noise experiments could then be used to measure the magnitude

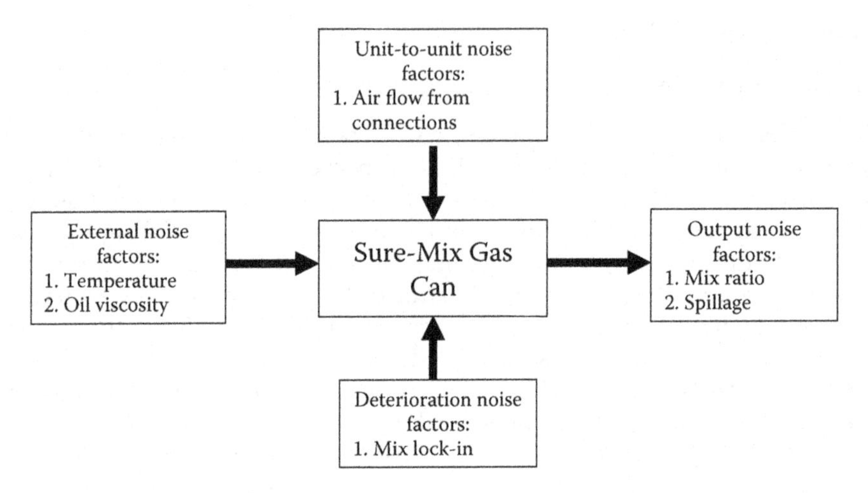

FIGURE 24.15
P-diagram.

of the impact of each of the noise factors. By doing this, it would be possible to determine whether or not the noises were significant, as well as to identify and prioritize the noises that would have the greatest impact on the design. Further, compounded effects, such as temperature and type of oil, could also be explored through experiments. Following these results, control factors would be implemented, and experiments for robustness optimization would take place. An example of a control factor on the gas can would be to have two settings (high and low temperature) for each mixing ratio to reduce the inaccuracy of the fuel to oil ratio poured. This would then be tested to statistically determine its effects on the robustness, resulting in signal-to-noise (S/N) metrics, which represent the gains in robustness and can be used in the additive model.

System Additive Model

Following robustness experiments, the S/N gains would be statistically proved to a degree of certainty through repetition of tests. From this, the impact of each of the control factors would be represented by a decibel quantity. To represent the additive model, Equation 24.1 was used.

$$S/N_{predicted} = S/N_{avg} + (S/N_A - S/N_{avg}) + (S/N_B - S/N_{avg}) + \cdots (S/N_A - S/N_{avg}) \tag{24.1}$$

With this model, A and B would be control factors, and the S/N would be the results when implemented and the average when not. This additive model would then be used to represent the overall improvement in reducing the variation of the CFRs. It is also important to note that control factors that do not have a large impact are not included in this calculation. For the CFR of variable selectable mechanism measure through accuracy, control factors would be the dual settings and increasing the tab that locks the selector in the correct position. These two control factors would reduce the variation due to noise and increase the robustness in this area.

Variational Sensitivities and System Variance Model

Across the technology platform, several variations are seen. The flow rates of oil and gas are among the major variational sensitivities. Although the flow rate of oil is regulated for a given mix ratio, variation can be seen. The type of valve used and the appropriate amount of opening are also considered major variational sensitivities. The type of valve used is a major concern when deciding the oil and gas flow rates. Also, the flow rate depends on the viscosity of the oil. A grade of oil with a particular viscosity should be suggested for use with the gas can. Thus, the system variance model can be given as in Equation 24.2.

$$\sigma^2_{\text{total}} = \sigma^2_{\text{gas flow rate}} + \sigma^2_{\text{oil flow rate}} + \sigma^2_{\text{viscosity}} + \sigma^2_{\text{valve}} \tag{24.2}$$

Overall, the system is robust, and only minor variances can be noticed.

Customer Reviews

After designing the product to meet the customer requirements, the team sent out the design to several customers for feedback. The team received a few suggestions on the technical aspects, including

- "I'm curious if the fins would cause turbulence in the flow causing gas to splash in an unpredictable manner when exiting the spout—many existing products already have this problem (possibly caused by the bellow shaped fins). Also, I tend to like vented gas cans better as they are less prone to splashing—adding a vent could be a feature you could implement."
- "Assuming that the graded mixing valve can accurately regulate the fuel/oil reasonably all the key elements are in place for the idea to work. I would recommend that a way to regulate the flow of fuel

also be incorporated in the design, perhaps a simple orifice for the gas flow. This would allow for better control of the mixing through a greater range of pouring angles for the gas can."

The following are some additional responses:

- Willing to spend $5–$10 extra.
- It solves a real-world problem.
- I think I would buy; however, I do not currently have a need for this product.
- $3 extra.
- Overcame the general difficulties faced in a general gas can.
- I have no changes to suggest but would like to see a valid working prototype assuring it is working.
- Yes, I would definitely give it a try.
- I think the idea of the redesign gas can is a good one.
- Perhaps up to 125% of the purchase price of an ordinary gas can.
- I like the visibility of fluids, convenience, and accuracy of mixing.

Lessons Learned

Part of the reason for choosing to use a PERT chart instead of Microsoft Project was the ability to incorporate the product development tollgates/phase gate system. The tollgate system provided an initial jumping-off point in the early stages of the team's work distribution and meeting schedule, listed as tasks on the PERT chart. This proved to be another tool in the DFSS process that, although not required for this report, was beneficial.

For this reason, the team decided to include an additional section for lessons learned and future planning, as prescribed by the tollgate system (tollgate #8), and will cover the areas of project review for future targets, field testing, and a brief summary of the project. Because the tollgate system is designed for projects that have been taken to the manufacturing and distribution phases, a modified take on these areas is presented.

Future Project Targets

The advantage of the final design that was produced through DFSS is the relatively small change that needs to be made to existing gas can manufacturing processes. The only real change to existing gas can designs is in the oil reservoir, which should not require much alteration in already established molds. The target analysis would need to focus on ease and cost of changing existing gas can mold manufacturing techniques as well as factory owners' level of interest. The Sure Mix Gas Can spout will need to be more fully developed. The future target should be a prototype design centered on mathematical design and experimentation in terms of totality of gas/oil mixing, universality, and cost reduction.

Field Testing (Prototype Acceptance)

Although the design process ended with the prototype, a modified field test was conducted in the form of several focus groups. Using prototype pictures sent to individuals and in-person feedback forums, several lessons learned can be derived that provide a guide for future marketing and further designs.

Two points were uncovered during the focus group/field testing of the prototype. First, the acceptance of the prototype was higher in rural areas, where people had a higher propensity to have multiple types of 2-cycle engines and were more prone to do their own lawn work or use recreational vehicles that required a gas/oil mix. Second, businesses that use gas/oil mixtures frequently, watercraft, lawn care, and go-carts, among others, were very receptive to the prototype but required a capacity of no less than 30 gallons. Both points demonstrate that DFSS was successful in meeting the quality, performance, and satisfaction that the customer required; however, consideration should be given moving forward to marketing to locations and groups that will be more likely to buy the product. Additionally, it may be beneficial to perform another DFSS project for the modification of the Sure Mix 5-gallon gas can into a vessel of higher capacity.

Summary

The beauty of this project was the implementation of DFSS concepts into the product design in almost real time. This allowed the team to take the concepts from class and put them into action, culminating in a prototype and analysis that provide an extremely well-rounded understanding of how DFSS guides idea generation, reduces different problems that can be encountered in the design process, and ultimately makes the design a reality.

Conclusion

The team feels confident in the results achieved through using the structured approach of DFSS and its tools. DFSS has allowed a successful modification of an existing 5-gallon gas can and an exciting innovation/invention in the arena of spout mixing technology. The main deliverables of the Sure Mix Gas Can are contained in its ability to mix oil and gas at any one of the five types of gas/oil mix ratios needed for standard 2-stroke engines while minimizing human behavior change. Moreover, the guiding principles for the project, presented in the charter at the start of the project, which stressed convenience to the customer and the reduction of negative environmental impacts while remaining financially feasible, were met.

25

Design for Six Sigma Case Study: Portable Energy Solutions

Tim Clarke, Matthew Fletcher, Eric Sears, Tim Wu, and Elizabeth Cudney

The Design for Six Sigma (DFSS) methodology enables a development team to achieve the necessary goals for the success of their product. It enables them to conceive new product requirements and systems based on a balance between the voice of the process and the voice of the customer (VOC). It also allows them to design baselines that are able to satisfy product requirements, based on the same balance. It further enables teams to optimize the performance of the design process to incorporate tunability and robustness into the product and fortify it against everyday sources of variation. DFSS also allows teams to verify how well their design holds up to industry standard by offering multiple testing methods that enable the team to measure how much variation their design contains.

Project Description

This project focused on creating a portable charging solution for everyday consumers who need to charge their phone on the go or take long trips in places that do not have wall chargers, such as the beach or a bike trail. An inherent aspect of the project was the need to incorporate previously unexploited sources of portable energy to increase the versatility of the product. To ascertain these sources, the team considered the options currently technologically available, including external lithium batteries, alkaline (AAA) batteries, solar panels, mini wind turbines, solar heating dishes, and portable charging solutions, and multiple combinations thereof. The team constructed an extensive survey to gather the VOC to form a valid sample. From the survey, the team defined, structured, and ranked the VOC to develop the product. Kano analysis helped break customer expectations into one-dimensional, expected, and exciting qualities. Using quality function deployment (QFD), the team integrated customer requirements into the design through functional requirements. The VOC was then validated, prioritized, and benchmarked in the house of quality (HOQ). Competitive evaluations of existing products on the market showed the existing benchmark and where there were opportunities for improvement. Using the design for X methodology, criteria were developed to formulate the basis of the conceptual designs. Seven design concepts were generated that met the customer's requirements from the HOQ. The team conducted Pugh's concept selection matrix to converge on a superior concept by evaluating the various concepts to a best-in-class benchmarked datum concept (super concept). From there, tools were used to minimize noise, improve reliability and function, and prevent failure in the device included modeling of technology, modeling of the robustness and tunability of the mean critical functional responses, the system additive model, modeling of vibrational sensitivities across the integrated technology platform, the system variance model, and robustness additive models.

Project Goals

The project goals were relatively simple but challenging. The team wanted to incorporate the VOC as much as possible into the design process, and during our surveys we encountered many opinions that would challenge us to expand our horizons. For instance, this new cell phone energy source had to be light and small enough for a person to carry around while it was attached to the phone and running. Current battery technology has recently reached a point of compactness that would allow us to achieve this goal with combinations of lithium batteries and green energy sources.

Requirements and Expectations

The requirements and expectations were partially developed by the customer, as any design that properly incorporates the VOC should be. The product needed to charge a variety of devices (iPads, Kindles, cell phones, etc.) with a single charging system. This meant 1 and 2 A variable charging, with a minimum 5 V output. The system needed to be robust and long-lasting; therefore, a minimum 5000 mAh was required to enable repeatable, speedy recharges.

Project Boundaries

The project would be able to charge small, portable devices; however, designing the systems for characteristics specifically found in laptops was outside the scope. While the device would certainly be able to charge a laptop, given its 1 and 2 A variable output and lithium battery system, the team did not incorporate laptop considerations into the design process, because customers were mainly looking for a charging solution for everyday, on-the-go users.

Project Management

The invent and innovate, develop, optimize, and verify (I²DOV) roadmap consists of four phases: invent and innovate phase and gate, develop phase and gate, optimize phase and gate, and verify phase and gate. During the first phase, the team defined the business goals, business strategy, and mission. Markets and market segments were also defined: everyday consumers with smartphones. The team analyzed the technology roadmap of cellular and tablet technology and realized that they were only becoming more popular, and would most likely continue to increase in popularity for a considerable period of time.

During the second phase, the team analyzed what superior systems concepts were viable for production. Seven prototype models were developed to identify what attributes would be desirable in a full-scale production model, each with realistic and unique characteristics. Pugh's concept selection matrix was used to isolate the superior characteristics. In the third phase, a superior concept was created to fulfill the VOC and the voice of the process, thus optimizing the product. In the fourth and final phase, design failure mode and effects analysis (DFMEA) was conducted to verify the success of the final product.

Invent/Innovate

Because the goal of the project was the commercialization of a product, the team needed to clearly understand two things: the VOC and the voice of technology. The VOC is what your customers need, and the voice of technology is what product and manufacturing technology is currently capable of providing. Understanding what customers wanted and needed was critical to the design development. To gather the VOC, a survey was created. First, the market segment was defined as smartphone users. A representative sample was surveyed that provided an accurate portrayal of the smartphone market, where the majority of customers are 22–35 years of age. The survey questions were tailored around the usage time that smartphone users spend on their phones, the applications they ran, recharge times, and frequency of battery shortages. The participants were also asked what desired qualities in a portable charging device were of interest. Using an online survey platform, the team used social media as well as e-mail to gather 80 surveys. The survey results are shown in Figure 25.1.

FIGURE 25.1
Survey responses.

About 94% of survey respondents were smartphone users, who were our target demographic. Almost 40% of those surveyed used a phone for more than 10 hours between charges, which indicated that there was potentially a need for a portable charger independent of a wall outlet. The survey results also indicated that smartphone users frequently used voice, text, Internet, gaming, and global positioning system (GPS) applications on their phones, which could quickly drain the battery. In addition, 30% of those surveyed had their battery completely depleted two to three times a month, which, combined with a response from 94% of those surveyed owning car chargers, further validated the need for charging capability outside the home. Finally, 87% of those surveyed indicated that they would be interested in a portable phone charger, which indicated that our product had potential for a niche in the market. To study what customers wanted to see in the product, participants were asked to rank important features. In order of preference, customers valued ease of use, size, ability to detach from the phone when not in use, universal compatibility between brands, and permanent connection to the phone. In terms of performance, about 44% desired charging capability similar to alternating current (ac) charging out of a wall outlet, 41% wanted a charge similar to USB charging from a laptop, 11% were fine with emergency charging only, and fewer than 4% wanted supplementary charging that slows the rate at which power is consumed. Of those surveyed, 53% were willing to pay $10 to $30 for a portable charger, while 35% were willing to spend $30 to $50. Outliers of approximately 5% on each end wanted to spend less than $10 or more than $50. Finally, survey respondents were asked to write in their own words the characteristics that they would like to see in this type of product. The responses received included size, performance, universal compatibility, durability, simplicity, retractable cord, cost, colors, safety, reliability, and compatibility with other devices such as Nintendo and iPads. The frequency with which these characteristics appeared enabled the team to rank them in order of importance to the customer.

Once a sufficient pool of responses to draw data from was reached, the team began to define, structure, and rank the VOC for the development of the new technology. The team wanted to define our technology requirements from customer needs that are new, unique, and difficult (NUD). Based on the survey results, the VOC was prioritized by assigning different weights to the characteristics. The VOC represented customer requirements, or the "what" in what had to be satisfied. This went into the left column of the system-level HOQ. The team focused on needs that were new, unique, or difficult to fulfill. In order of importance (weight), the customer requirements were simplicity, size, rate of charge, charge capacity, reliability, universal compatibility, durability, cost, useful life, and aesthetics. While most of these characteristics were not new, unique, or difficult on their own, they could be combined into one synergized product that would be new and unique. The difficulty lay in determining how to do it, which led the team to QFD.

Quality Function Deployment (QFD)

QFD is a systematic process to integrate customer requirements into every aspect of the design and delivery of products and services. QFD is a collection of matrices that are used to facilitate group discussions and decision-making. The HOQ in QFD was used to determine the design requirements (how to satisfy the customer based on the VOC) by fulfilling needs through the functional requirements of the product (Figure 25.2). The design requirements that were developed to meet the needs of the customer were USB interface (universal compatibility), size dimensions less than 4″ × 2″ × 0.5″ (size similar to iPhone), 5 V output and 1 A current

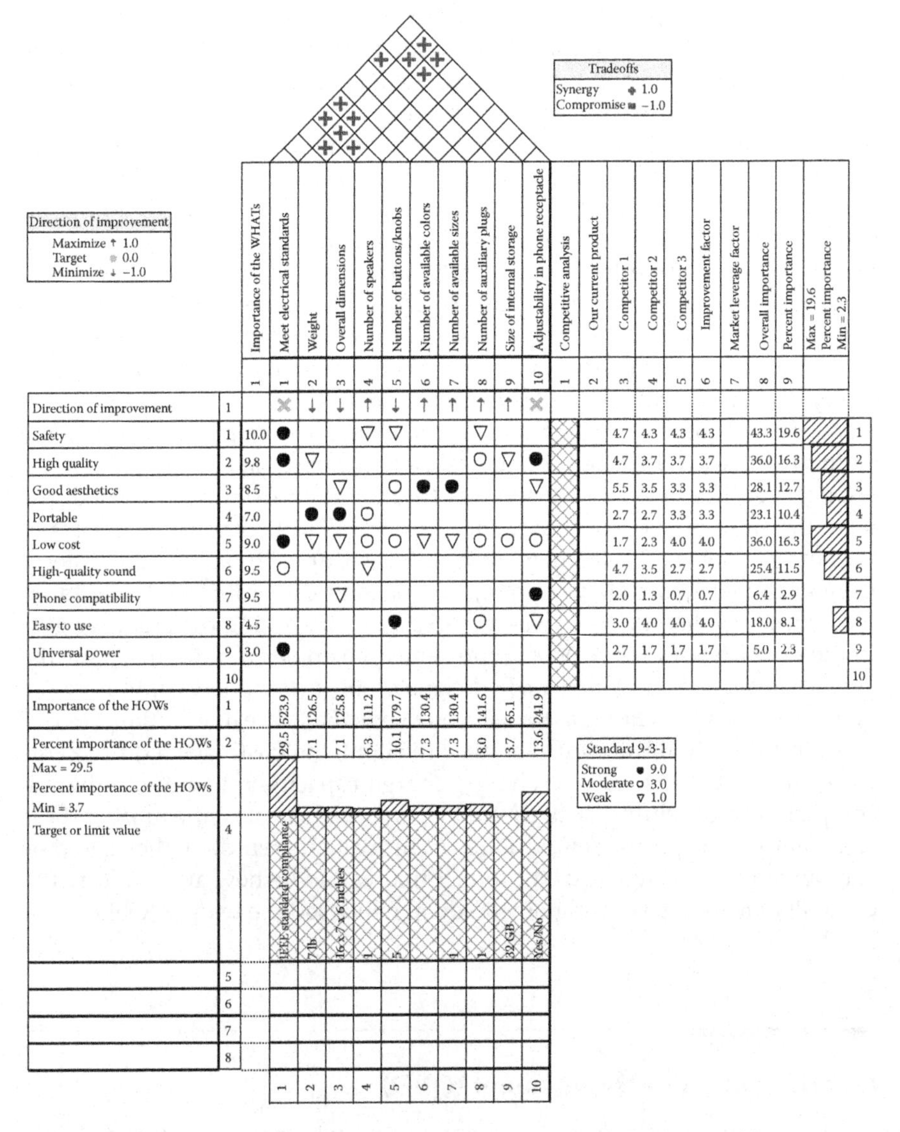

FIGURE 25.2
Quality Function Deployment matrix.

(performance similar to wall outlet), $50 maximum (price), the ability to withstand a 3′ drop (durability), minimum capacity of 5000 mAh (performance), environmentally robust (performance, durability), color availability (aesthetics), plug and play (simplicity), status indicator (delighter), less than 95% loss in power over time (performance), and sleek and clean appearance (aesthetics). Through QFD, the team gained a better understanding of customer requirements that could be used to increase customer satisfaction while reducing time to market and decreasing development costs. It increased the team's ability to create innovative design solutions and enhanced our capability to identify specific design aspects that had the greatest overall impact on customer satisfaction.

The HOQ enabled the comparison of customer requirements (VOC) with design requirements. Based on how the VOC was weighted, the team determined that price, size, universal compatibility, charging performance, and capability were the key requirements to satisfying the VOC. There were also strong correlations between size, cost, and performance. The main constraints in the design came from cost and size, while a USB interface and performance measures similar to those of a wall charger satisfied many of the customers' needs. Lastly, a comparison with competitors' products indicated what could be improved in the current market. Many of the products had some elements of our key requirements; for example, a product might have good performance but at the expense of cost and size. Another product might be compact and inexpensive but sacrifice performance. The team was able to discover a niche in the market by seeing how the competition performed against our requirements and potential solutions.

With Phase I of the product planning phase complete, the team continued QFD through the four-phase approach. Phase II, the part deployment phase, uses the prioritized design requirements to identify key part characteristics. In Phase III, the process planning phase, key parts are used to systematically identify process operations critical to ensuring customer requirements. Phase IV, the production planning phase, uses specific production requirements identified to ensure that key processes from Phase III are controlled and maintained.

Kano Analysis

Kano analysis (shown in Figure 25.3) assists in processing the VOC data. The Kano Model defines three types of quality requirements: one-dimensional, expected, and exciting qualities. The Kano Model also defines how the achievement of these requirements affects customer satisfaction. One-dimensional qualities are specifically requested items that customers stated

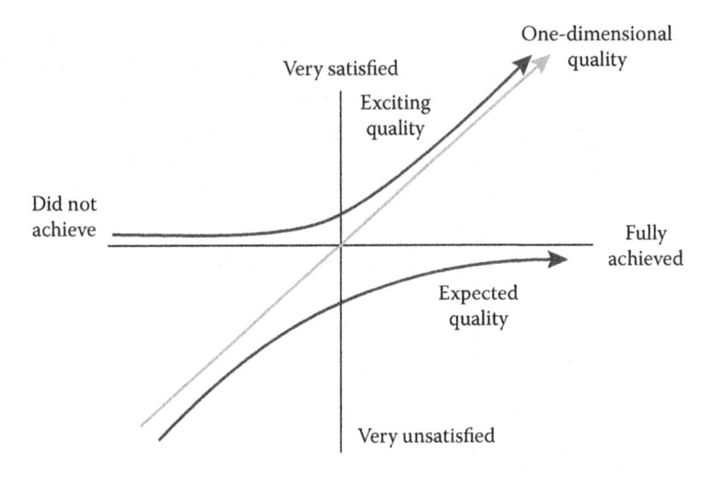

FIGURE 25.3
Kano Model.

in the survey. If these are present, the customer is satisfied; if they are absent, the customer is dissatisfied. Expected qualities are not specifically requested but are assumed by the customer to be present. If they are present, the customer is neither satisfied nor dissatisfied, but if they are absent, the customer is very dissatisfied. Exciting qualities are unknown to the customer and are the most difficult to define and develop. If they are present, the customer is very satisfied, and if they are absent, the customer is neither satisfied nor dissatisfied.

Most of our survey questions focused on one-dimensional qualities, such as performance measures, size, capabilities, compatibility, and cost. Expected qualities included characteristics that the customer may take for granted, such as durability, reliability, safety, robustness in different environmental conditions, and durability. The team suspected that our customers would treat a portable charger like a cell phone, and that our product would need to withstand everything a cell phone could. This gave the team a realistic benchmark for comparison. The product would be expected to work every time it was used, and should be able to hold a charge for at least a month. Parts should not break easily, and users should not be concerned for their safety when using the charging device. The team was able to find exciting qualities by offering status indicator lights. While this was not explicitly stated as a need by most users, we predicted that LED lights indicating the charging status of the device would be appreciated and would allow users to know the condition of their charger from a quick glance. Other potential exciting qualities came from color availability and a sleek and clean design. By offering a visually appealing product customizable to user tastes, the team hoped to set the product apart in the market.

Develop

Design for X Methods

Prior to generating the seven concepts, the team developed criteria to formulate the basis for the designs. These criteria were to design for manufacturability, reliability, serviceability, testability, and the environment.

When designing for manufacturability, the goal was to avoid complexity and produce easy-to-manufacture products. This not only simplifies the process and potentially produces fewer defects, but also theoretically reduces the cost to manufacture. Designing for reliability was aimed at a robust design that minimizes noise factors. Examples of this include reverse engineering the leading industry products outlined in the competitive analysis in the HOQ; using empirical tolerance designs, as in the power output for a given size of solar panel or lithium ion battery; and simplifying the parts and interfaces of each design. Designing for the environment does not refer to using recyclable materials or being environmentally friendly. The intention was that this charging device would be disposed of in the same way as a cell phone. Design for the environment referred to the robustness of the designs in various environments such as heat, sunlight, water, and dust. The standard chosen was that the concept should be able to withstand the same environmental impacts as a typical cell phone. Testability was introduced later, after the team had conducted a DFMEA and used inspections and testing at the end of the production line to mitigate the risk associated with the super concept. Examples of the testing introduced include the testing of USB ports, solar panels, batteries, and indicator lights. This consideration was retroactively applied to the concepts, but the emphasis remained on the super concept.

Concept Generation

Using the information gathered up to this point, concept generation was initiated, with each team member designing one or two concepts. General guidelines were discussed as a group to ensure that duplicate concepts were not produced. The first three concepts are largely based on products already available on the market with minor changes. The final four concepts are not currently available on the market and are a blend of existing technologies adapted for our purpose. The seven concepts are

1. **AAA battery phone charger (Figure 25.4)**: This concept was derived from products that already exist in the market and were identified in the competitive analysis. The advantages of this concept are simplicity, reusability, price, and size, while a disadvantage is the lower performance in terms of power output.

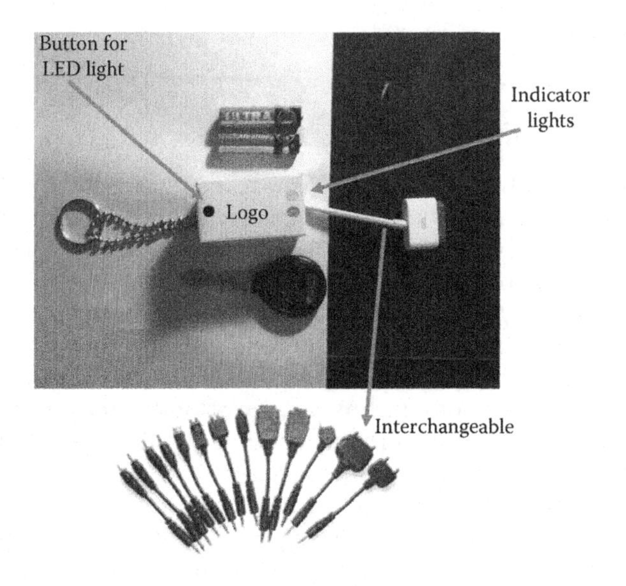

FIGURE 25.4
Concept 1.

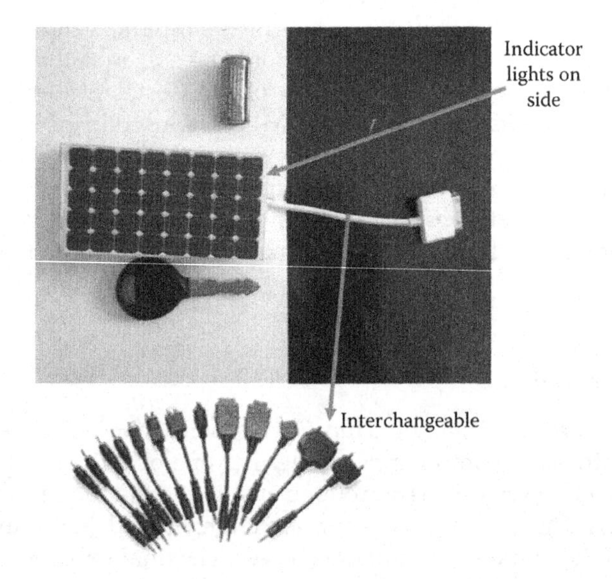

FIGURE 25.5
Concept 2.

2. **Solar and lithium ion phone charger (Figure 25.5)**: A combination of a powerful lithium ion battery and the longevity of a solar panel are the key features of this concept. The benefits include a powerful charger that exceeds every performance measure regarding power delivery in addition to a residual charge provided by the solar panel.

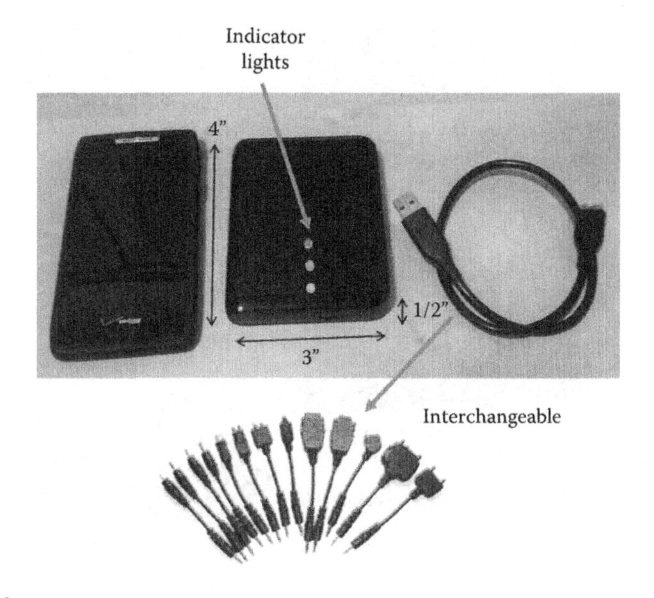

FIGURE 25.6
Concept 3.

While this concept is larger than some of the others, it is still within the functional requirements outlined in the HOQ and just below the price ceiling.

3. **Lithium ion phone charger (Figure 25.6)**: This concept is also based on similar products from the competitive analysis. The addition of indicator lights and dual USB ports differentiates this concept from existing products. These features allow the user to know the status of their charger and to charge their phone and tablet simultaneously. Similarly to Concept 2, this design is capable of providing the maximum power to safely charge a device, but is larger and heavier, and requires six hours to recharge to its full potential.
4. **Foldable solar panel (Figure 25.7)**: This concept uses a larger (12″×8″) solar panel that is made of pliable material that can be folded when not in use. The larger surface area allows improved performance, and this concept is able to operate independently of recharging or replacing batteries. However, these benefits come at a price over twice the established limit, the size is too large, and there is no ability to store power, since it is without capacitors or batteries.
5. **Solar-kinetic charger (Figure 25.8)**: This concept was designed to be attached to the back of the phone or its case semipermanently. The two sources of power are a solar surface and a kinetic generator. The hemispherical shape is covered with solar panels that provide a small amount of power from the 2″ diameter. The primary source of

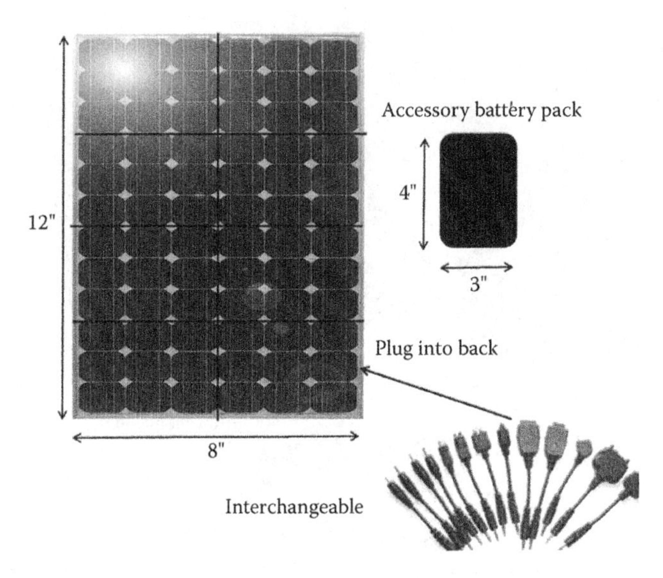

FIGURE 25.7
Concept 4.

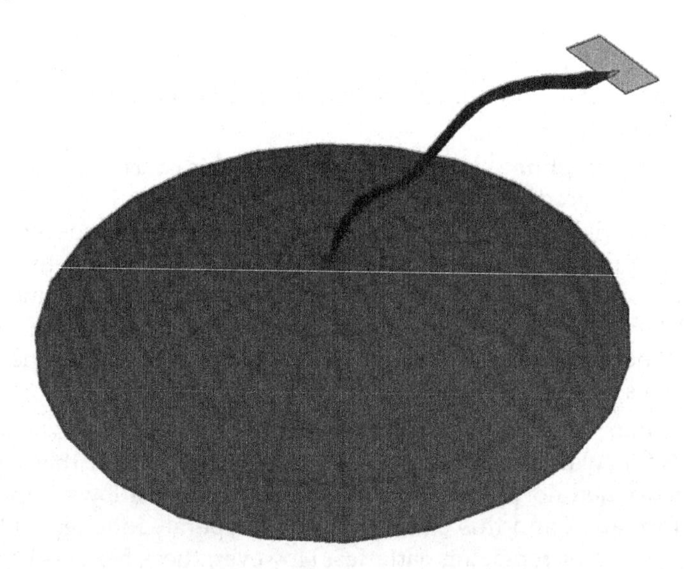

FIGURE 25.8
Concept 5.

power is the kinetic generator that is in the bottom half of the design. The premise is to use the same technology as is used in watches, which transfers movement into energy through a rotating pendulum that is attached to a large gear spinning at up to 100,000 rpm. The design also incorporates a retractable cord in the top half of the hemisphere, which transfers the power to the cell phone. The theory

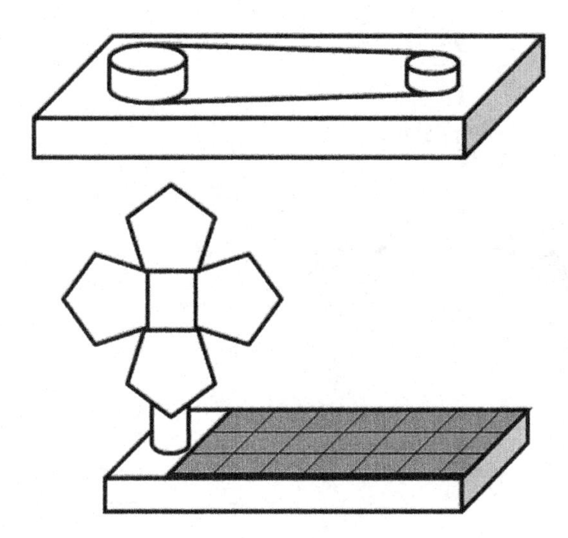

FIGURE 25.9
Concept 6.

behind this concept is that the user charges his or her phone while on the go. While the idea of portable reusable energy is possible, this concept would be very expensive and less powerful than many of the other options.

6. **The "MacGyver" (Figure 25.9)**: A concept that is geared toward the outdoorsman and emergencies, the MacGyver has it all. It combines the use of wind power, solar energy, lithium ion batteries, and even a wind-up hand crank. The benefits include quadruple redundancy in power sources and sufficient power output, primarily due to the lithium ion battery. The downsides are that this design is too large and too expensive to meet our initial design requirements, and using it can be quite complicated.
7. **Solar heating dish (Figure 25.10)**: The final concept uses another exotic technology, in which a highly polished concave mirror concentrates heat onto a Stirling engine at its center. The Stirling engine gyrates when heat is applied, and this friction is converted into electricity. Although this concept uses the sun, it is different from solar energy in that thermal energy does not absorb energy into photo cells. Instead, the mirrors generate the heat used to power the Stirling engine. The advantages include a renewable power source that can be used even when it is cloudy. The disadvantages include a complex, expensive design that is difficult to set up and use, because the dish must be at the correct angle, similarly to a satellite television dish.

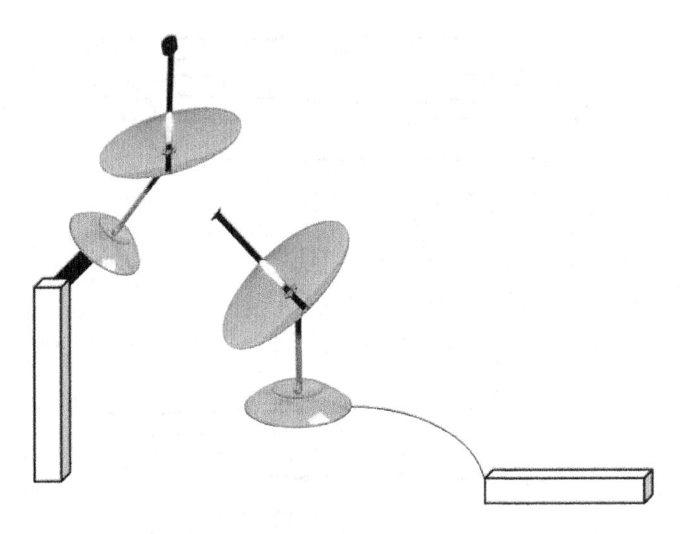

FIGURE 25.10
Concept 7.

Modeling of Technology

The ideal function for our product is quite simple with a single input and output. The signal (input) is the power generation from the given concept. The performance measure is the transfer of adequate energy to charge the cell phone. Based on our research, the intended result is to provide 5 V and 5000 mAh (±500). Unintended results include, but are not limited to, a failure to charge, overheating, or degrading performance.

In the P-diagram (Figure 25.11), the ideal function lies along the horizontal axis. The noise factors are on the top portion of the diagram and consist of conditions such as the environment, customer (mis)handling, and variation within our processes or production. Below the diagram are the control

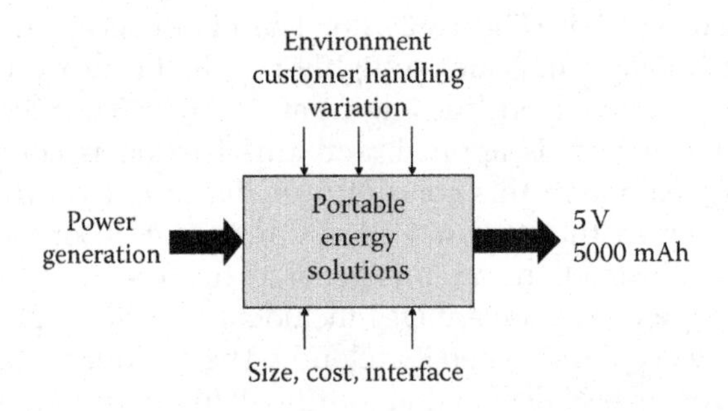

FIGURE 25.11
P-diagram.

factors, which are specific, are easy to manage, and have a big effect on performance. Control factors for our concepts include size (<4″×2″×½″), cost (<$50), interface (USB), and other scale factors that come from our functional requirements.

Super Concept

The super concept was based on the second concept, the solar and lithium ion phone charger. The super concept is able to charge a cell phone, GPS, and laptop with the option/combination of solar panel and lithium ion battery (shown in Figure 25.12). The combination of solar panel and lithium ion battery was selected because it would be ideal during power outages, camping, hiking, or sailing, or whenever emergency situations arose. The lithium ion battery was designed to be the main source of power for recharging the phone, while the solar panel acted as a supplemental or second power source. With the dual power source option, it was designed for the user to be able to make and receive calls as soon as it was plugged in. The super concept came with 10 charging adapters, which made it compatible with most cell phones. It was 4″ long and ½″ high and weighed less than 8 oz. Indicator lights were designed to show power and charging status. The protective case was designed to shield the solar panel, the lithium ion batteries, the indicator lights, and the USB port from damage. The price for the super concept came in at $49.99.

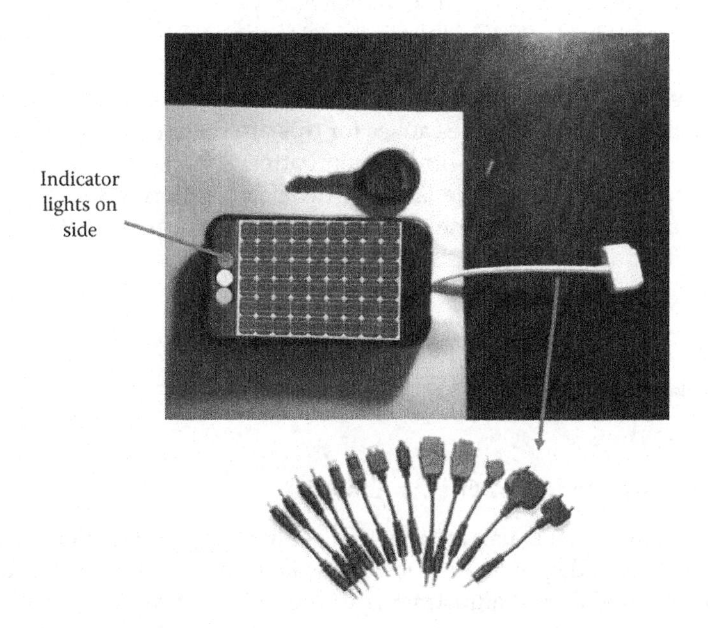

FIGURE 25.12
Final concept.

Pugh's Concept Selection

Pugh's concept selection matrix is a structured concept selection process used to converge on superior concepts by using a matrix consisting of criteria based on the VOC and its relationship to specific, candidate design concepts. Pugh's concept selection matrix was employed to develop characteristics for a super concept (Figure 25.13). The VOC was compared for each concept against a datum.

Optimize

Modeling of Robustness

Functional analysis system technique (FAST) trees graphically represent how each concept uses its technology to charge a cell phone. The FAST diagram shown in Figure 25.14 is based on our super concept, which has the sole purpose of charging a cell phone at the system level. At the subsystem level, two primary functions are occurring: generate power and transfer energy. In the super concept, power generation is achieved through a lithium ion battery and a solar panel at the subassembly and/or component levels. Energy is transferred via a USB interface, and a critical function is to regulate the power distribution at specified metrics (i.e., 5 V).

A fault tree (Figure 25.15) is the inverse of the function tree diagram and was used when developing the DFMEA to improve the robustness and reduce the noise of the super concept. An example of a fault is that our phone does not charge. Two possible causes for this are that no power is being generated by the solar panel or lithium ion battery, or that no energy is being transferred to the phone. The lack of power generation is self-explanatory, but the inability to transfer energy may stem from a broken USB port, failure in a circuit, or any number of issues.

Verify

Design Failure Mode and Effects Analysis

DFMEA is a tool that is used to identify, evaluate, and solve potential causes of failures before they happen. DFMEA was used in this project as a tool to validate and make final adjustments to the super concept. The DFMEA for the product was broken up into 12 categories: the specific item of the system, potential failure mode, potential effects of failure, severity rating, potential

	Concept solar + Li	Concept AAA	Concept foldable solar	Concept Li	Concept solar + Kinetic	Concept MacGyver	Concept solar heat	Super concept based on solar + Li	Change description
Simple	S	+	−	+	S	−	−	S	
Size	+	+	−	S	S	−	−	+	
Rate of charge	+	−	−	S	−	+	−	+	
Charge capacity	+	−	−	S	−	+	−	+	
Reliability	+	S	−	S	S	+	−	+	
Universal	S	S	S	+	S	S	S	New value: +	Dual amperage
Durable	−	S	−	S	−	−	−	New value: +	Design a case around the battery and solar panels
Cost	−	+	−	S	−	−	−	New value: S	Use smaller battery
Useful life	+	−	+	S	+	+	+	+	
Aesthetics	S	+	−	S	S	−	−	S	
Sum of (+)	5	4	1	2	1	4	1		
Sum of (−)	2	3	8	0	4	5	8		
Sum of (S)	3	3	1	8	5	1	1		

FIGURE 25.13
Pugh's concept selection matrix.

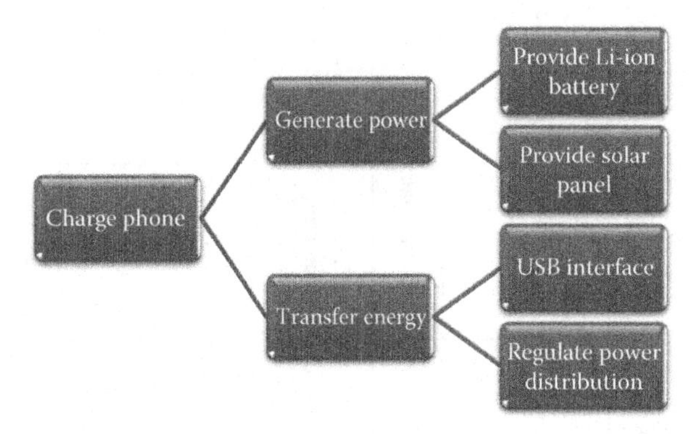

FIGURE 25.14
FAST diagram.

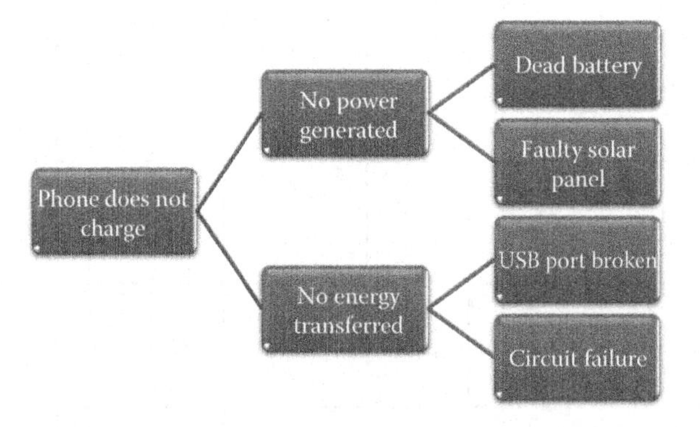

FIGURE 25.15
Fault tree diagram.

causes/mechanism of failure, occurrence rating, current design controls, detection rating, risk priority numbers (RPN), recommended actions, responsibility and target completion date, and the action results. The RPN in the DFMEA is a mathematical product that takes into account the severity, occurrence, and detection ratings. These ratings are based on a scale from 1 to 10. The severity scale uses 1 to illustrate that the item or function has no effect, whereas 10 could have hazardous implications for the user. The occurrence scale uses 1 to represent that it is unlikely that the failure will occur, whereas 10 indicates that the failure is almost inevitable. The detection scale uses 1 to illustrate that the design controls are almost certain to catch the potential fault or failure, but 10 represents absolute uncertainty that the problem will be detected. The end goal for the RPN is to have a lower value, due to improving design controls, than when you first addressed and identified the potential failure. For the portable energy source, four items/functions were identified that could lead to potential failures: the solar panel, the lithium ion battery, the USB port, and the indicator lights, as shown in Figure 25.16.

Potential failure mode and effects analysis

Design FMEA

Item: Sol-ion
Technology: Solar and Lithium Ion
Core Team: Bonobos
Design Responsibility:
Key Date: 02 MAY
FMEA Number: 1
Page 1 of 1
Prepared by: Team 2
FMEA date (Orig): 16 Apr

Item / Function	Potential Failure Mode	Potential Effect(s) of failure	SEV	Potential Cause(s)/ Mechanism(s) of failure	Occ.	Current Design Controls	Detec.	RPN	Recommended actions	Responsibility and target Completion date	Action results				
											Actions taken	Sev	Occ	Detec	RPN
Solar panel	Cracking, deformation	Inoperability, fragmentation of panel	7	Mishandling, excessive force	4	Addition of impact-resistant case	1	28	Visually inspect	Quality control of each solar panel before shipment	Inspection station at end of production line	7	3	1	21
Lithium ion battery	Enviornmental damage, leaks, battery fatigue	Inoperability, burns, fire, degraded performance, odors, destruction of property	9	Excessive heat, chemical failure, harsh operating conditions	3	Case	5	135	Fireproof case	Ensure UL standards are met	Test each battery	7	2	3	42
USB port	Breaking, shorting, excessive wear	Inoperability, degraded performance	6	Water damage, mishandling, rough usage	4	None	1	24	Rubber USB cover	Covers installed before shipment	Test each port	6	3	1	18
Case	Cracking, overheating	Less protection, degraded battery life	5	Wear and tear, high life	7	Material testing	6	210	Design case with ventilation in mind	Design teams conducts temperature experiments	Thermal scan at end of production line	5	4	4	80
Indicator lights	Light burnout, cracking	Loss of indication of charging status	4	Usage, stress or shock	2	LEDs	1	8	Recessed within case	Integrate into design	Test each light	4	1	1	4

FIGURE 25.16
FMEA.

Each of these item/function failures could be associated with some sort of cracking and deformation of the component. The potential effects of the four item failures were associated with inoperability and degraded performance. The recommended actions and design controls led to the development of an impact- and fire-resistant case that provided ventilation. This recommended action and design control led to the reduction of each item's RPN. Another item/function evaluated was the case itself. The case failure was also associated with some sort of cracking and deformation. The potential effects of the case failure were associated with less protection and possible degraded battery life. The recommended actions and design controls led to the development of ventilation holes in the case and making sure to conduct material testing to prevent and control cracking in the case. The results led to a reduction in the RPN from 210 to 80.

System Variance Model

The system variance model in the verify portion of the project can be geared toward making final adjustments in quality by reducing the variation of the process and improving the robustness. We can use the variance model to check and validate our process by making sure our subsystem parts fall within the upper and lower specification limits. Once we have the variance of each subsystem, we can use the variance model to find the overall variance of the system, as shown in Equation 25.1.

$$\sigma_{\text{total}}^2 = \sigma_A^2 + \sigma_B^2 + \cdots + \sigma_N^2 \tag{25.1}$$

To apply this model to the portable energy source, we would need to validate and reconfirm the variances of the solar panel, battery system, case, USB port, indicator lights, and so on. Then we would add these variances using the variance model. This would let us know whether we were on target with our upper and lower specification limits or whether we needed to go back and improve a subsystem process.

Develop and Confirm Robustness Additive Models

In the verify phase of the I^{2}DOV process, the system needs to be evaluated for its robustness before it reaches the customers. In evaluating the system's robustness, we are also focusing on the capability of the product design and functional performance. In confirming the robustness additive model, the system is tested for sources of variation that can occur during customer use. These sources of variation include external variation; subsystem and subassembly functional variation due to assembly, service critical adjustment parameter adjustment variation; component variation; and deterioration variation.

- External variation: To test external variation factors, you must run the system robustness test using external sources of variation that could have an effect on the product. Testing our portable energy source to external variations would include exposing it to drastic temperature changes, humidity, altitude fluctuations, and moisture.
- Subsystem and subassembly functional variation due to assembly, service critical adjustment parameter adjustment variation: To test for this type of variation, you would need to alter the critical adjustment parameters off target to simulate an error in the assembly process of the product. Testing the portable energy source to this variation would include altering the critical adjustment parameters to test the effects on and the outcome of the product if it was not assembled correctly. An example of this variation on our portable energy source would be to broaden the parameters for installing the solar panel. After this has been done, the system variations should be tested to determine whether this has any effect on energy output.
- Component variation: To test for component variation, there must be component dimensions that are off target to simulate variation from the supply chain. Testing the portable energy source to component variation would include assembling components that were off by one standard deviation from their target requirements and testing their effects. An example of this process would be to acquire a frame, a solar panel, and a lithium battery that were off target and to test the variation due to each.
- Deterioration variation: To test for deterioration variation, the deterioration process must be simulated over a given time. The estimated deterioration process will eliminate surprises that potential customers might face with the product in the future. To conduct deterioration variation on the portable energy source, we would have to alter components to simulate a deterioration for a given time. The time frame would be one, two, and three years, since the average cell phone is used for two to three years. After the components had been altered to simulate the aging process for these periods, the variations and effects this had on the product would then need to be tested.

To validate and calculate the optimal level of system critical functional response (CFR) robustness for the portable energy source, the additive model shown in Equation 25.2 can be used.

$$S/N_{predicted} = S/N_{avg} + (S/N_A - S/N_{avg}) + (S/N_B - S/N_{avg}) \\ + \cdots (S/N_A - S/N_{avg}) \quad (25.2)$$

This will allow the team to validate the final design by confirming and optimizing the signal-to-noise ratio based on these variations in the portable energy source.

Conclusion

The DFSS process enabled the development of a portable charging device that met the needs of the customer. The team was able to meet the needs of the customer by deploying the VOC throughout the different phases of the product design. The DFSS process also enabled efficient execution of the development of the portable energy source and prevention of development rework. Throughout the development of the project, the best practices and tools from DFSS methodology were employed. The team constructed an extensive survey to gather the VOC. This survey allowed the group to understand the customer needs and desires regarding portable energy for cell phones. The VOC in the HOQ was then analyzed and compared with the functional design requirements. After the HOQ was defined, seven concepts were developed. Design for X helped define important factors that were applied to each concept. Modeling of technology, modeling of robustness and tunability of the mean CFRs, and the system additive model further refined the product development process. Pugh's concept selection matrix was conducted to narrow down the seven concepts to one super concept. The super concept was validated through the DFMEA process, which helped identify potential failures in the final product concept. The system variance model and robustness additive model allowed final adjustments to the product before production. The steps and tools involved in DFSS allowed the team to create a potential product that would avoid manufacturing and service process problems. At the end of the project, the team examined customer feedback to the super concept, which was positive. The universal compatibility and solar charging capability were key aspects of the product that were highlighted in the feedback. Some suggestions for improvement included the ability to attach the product to a cell phone as one unit, an accessory to attach it while running outside, and a retractable charging cord.

26

Design for Six Sigma Case Study: Paper Shredder

Keith Major, T.J. McHugh, and Elizabeth Cudney

Six Sigma is a process improvement philosophy that views all works as processes that can be defined, measured, analyzed, improved, and controlled through the use of problem-solving and statistical tools. The goal of Six Sigma is to prevent defects by eliminating variation and waste, while delivering value-added and quality products to the customer. Six Sigma methods originated at the Motorola Company and were further refined by Jack Welch on leaving the Motorola Company and becoming the chief executive officer of General Electric. Design for Six Sigma (DFSS) takes these basic principles and applies them to product, process, and service development. DFSS has a five-phase structured approach: define, measure, analyze, design, and verify. This method is used to prevent design problems instead of using additional resources to fix the problems in the future.

Project Description

This project will seek to redesign the standard paper shredder to improve the compaction and/or containment of the debris that is the result of the paper-shredding process based on feedback received from potential customers. When one attempts to empty a trash-can-size home or office paper shredder, it is common for the debris to create a mess that requires additional cleaning time. This mess is generally caused when (1) the trash can responsible for catching the paper shreds is removed from the paper shredder itself; (2) the shreds of paper are being transferred to a separate container for disposal purposes; or (3) the teeth of the shredder are being cleaned. The scope of the project will include personal home and office paper shredders that must be emptied into another container for disposal.

		February				March				April				May	
		Wk 1	Wk 2	Wk 3	Wk 4	Wk 1	Wk 2	Wk 3	Wk 4	Wk 1	Wk 2	Wk 3	Wk 4	Wk 1	Wk 2
1	Development of project charter	■	■												
2	Development of I2DOV roadmap			■	■										
3	Invent/Innovate Phase														
3.1	Develop survey questionnaire				■										
3.2	Conduct survey					■	■								
3.3	Analyze survey results							■							
3.4	Quality function deployment								■	■					
4	Development Phase														
4.1	Concept generation										■				
4.2	Pugh concept selection											■			
4.3	Develop FMEA on selected concept												■		
5	Design Optimization														
5.1	Prototype development												■		
6	Product Verification														
6.1	Modeling of variational sensitivities													■	
7	Prepare DFSS report													■	
8	Project wrap-up													■	
9	End of project														■

FIGURE 26.1
Project plan.

Project Goals and Requirements

The goal of this project is to provide individuals who use home or office paper shredders with a product that will reduce the amount of paper debris that is typically created on emptying or cleaning. The success of the product will be determined by the increase of market share and customer feedback.

Project Management

To remain on track for successful completion of this project, the team created a Gantt chart that outlined all of the necessary tasks for the project, as shown in Figure 26.1. The project plan gave the team the ability to quickly and easily see at a glance whether we were on track, behind, or ahead of the project plan. By reviewing this document weekly, we were able to discuss as a group what needed to be done to ensure that we remained on task, so that we would complete all of the project components as needed to have a successful project.

Invent/Innovate

Gathering the Voice of the Customer (VOC)

The team gathered the VOC by using an 11-question survey with questions on the demographics of the user, the type of shredder the users had experience with, and their feedback on features they liked in a shredder. We used the survey to determine whether there was a need for a "cleaner" paper shredder and also to gather user ideas about additional features that would make the paper shredder better. The survey was administered through an online link, and the results were stored in an Excel spreadsheet.

Voice of the Customer

The VOC was obtained through the analysis of the survey results. The survey responses were stored in an Excel document that enabled organization and analysis. There were a total of 152 survey respondents. To ensure that unbiased results were obtained, demographic questions were asked prior to the questions designed to identify the VOC. Of the 152 respondents, 53% were male and 47% were female. There were also five age group selections, with 26–35 having the largest number of respondents with 69, whereas 18–25 was second with 30. The majority of the respondents had experience using paper shredders in a personal or small office setting, which is the scope of our project. Additionally, 12 individuals who took the survey had never used a paper shredder.

After covering the demographics, the team was then able to understand the VOC. The survey included several questions that asked the survey taker to rank his or her response on a scale in ascending or descending order. These responses were analyzed by giving a higher rating to the more favorable responses and then summing the values across, as shown in Figure 26.2. In this example, responses in column 1 were given a weight of 5, and responses in column 5 were given a weight of 1.

In addition to the numerical response questions, the survey included questions that allowed the survey takers to write in their own responses. For these questions, affinity diagrams were used to determine the top responses and to group them together. Using these two analysis methods, the team was

Purpose of using a paper shredder?

	1	2	3	4	5	Score
Personal	114	0	0	0	0	570
Small office	15	20	13	0	0	194
Large office	11	19	6	6	0	161
Home office	7	36	0	0	0	179
Industrial	1	1	0	2	3	16

FIGURE 26.2
Summary of data on survey responses.

Ideal shredder
Safe operation
Self-cleaning teeth
Containment of shredded particles
Easy to empty
Overfill prevention
Compaction of particles
Single-hand usage
Less jamming

Causes for mess
Particles land everywhere
Paper overflows
Shreds caught in teeth
Removable bin

FIGURE 26.3
Affinity diagram of open responses.

able to clearly identify the VOC. Figure 26.3 lists the items that the respondents stated would make for the ideal paper shredder and the major causes of mess based on their responses.

An interesting insight that was identified from the survey results was that customers did not feel that emptying the paper shredder was as messy a process as the team had assumed prior to distributing the survey. On a scale of 1 to 10, with 10 being "Very Messy," the average response was 4.3. Another unexpected result was that the customers ranked emptying a paper shredder with an average response of 7.5, with 1 being "Difficult" and 10 being "Easy." These two responses resulted in the team slightly altering their initial approach to the project to make sure that we did not impose our own thoughts on customers but rather, that we listened to what the data was saying and developed concepts that satisfied their needs.

Quality Function Deployment (QFD)

The tool that we chose to use for the QFD process was the house of quality (HOQ). To use the HOQ, the team had to first understand the VOC and identify the specific features that the customer is willing to pay for in the ideal paper shredder. Some of those features were explicitly stated in the survey data received; other features were left unstated, and were simply expected and understood to be the norm. Using the survey response and comparable machines that are currently on the market today, the team was able to identify the explicit and implicit customer requirements. Those requirements are listed on the left hand side of Figure 26.4. To satisfy these customer requirements, we then had to outline functional requirements that had a direct correlation to one or more of these customer requirements. These functional

Customer requirements (explicit and implicit) \ Functional requirements	Power shutoff function	Forward/reverse option	Full bin indicator	Removable catch bin/shredder	Lightweight	Compactor mechanism	Paper sensor	Self-lubricating	Shredder containment system	Paper guide	Paper thickness check
Safe operation	●	○			▽		●			▽	▽
Self-cleaning teeth		●				▽		●			▽
Containment of shredded particles			▽	●		●			●		
Easy to empty			▽	○	●			▽	○		
Overfill prevention	○		●			○					
Compaction of particles			○			●		▽			
Single-hand usage					●		▽			○	▽
Less jamming								○		●	●

FIGURE 26.4
HOQ.

requirements are listed across the top of the HOQ center matrix. After the lists of customer and functional requirements were generated, the relationships between the two were categorized to ensure that all of the customer requirements would be satisfied.

The team next had to categorize the relationships between the functional requirements so that we were aware of any possible conflicts in functions before beginning the design phase. This was performed in the "roof" section of the HOQ, as shown in Figure 26.5. A + sign indicates a positive relationship between two functional requirements, while a – sign indicates a negative relationship between two functional requirements. The "direction of improvement" line indicates whether more or less of this functional requirement would improve the overall quality of the product. Some requirements need to meet a specific target, so they would not have a more or less designation assigned to them in those instances.

After completing the HOQ, the team was then ready to move into the conceptual design phase of the project. The team used the HOQ and the ranking of industry competitors to determine what we could use from industry and what we needed to develop ourselves. The entire HOQ is shown in Figure 26.6.

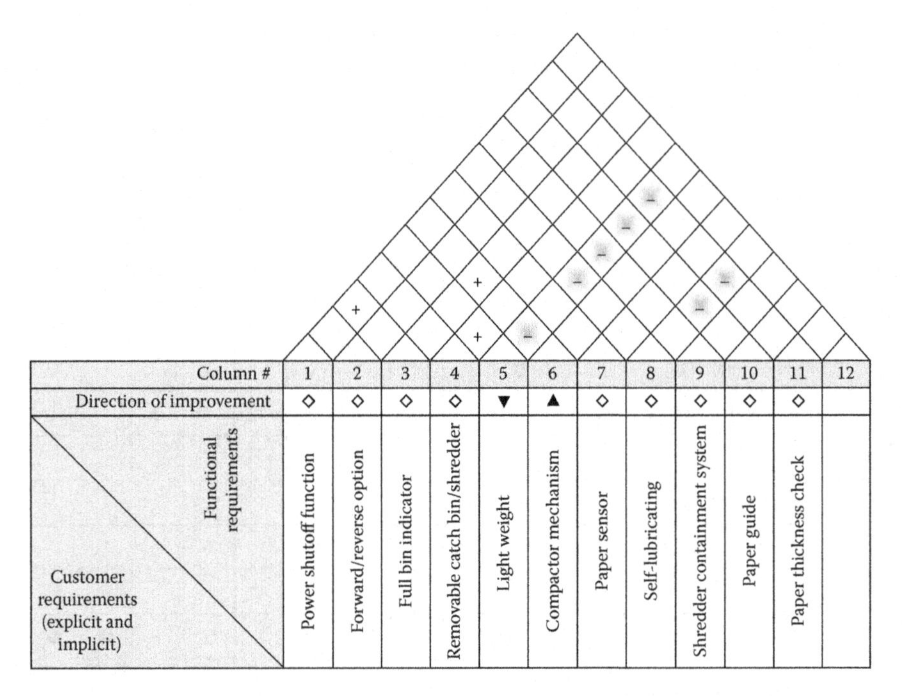

FIGURE 26.5
Roof of the HOQ.

Develop

Design for X Methods

When considering what concepts the team should develop, we needed to consider how our designs would be implemented in terms of manufacturing, servicing, and environment considerations. It is important to bring in a team of subject matter experts (SMEs) to determine methods for improving the manufacture, assembly, service, reliability, testing, and environmental impact of the proposed design. Some of the designs pushed the envelope to what is currently on the market; therefore, it would be imperative for the team to get SME feedback on our designs prior to selecting a final design. The team did consider the environmental impact of using plastic bags in the catch bin, and incorporated that in a couple of designs, but the final design left it out because of the additional environmental impact of using plastic bags.

Concept Generation

The team translated the customer requirements into functional requirements in the HOQ and built the designs to meet the customer requirements/

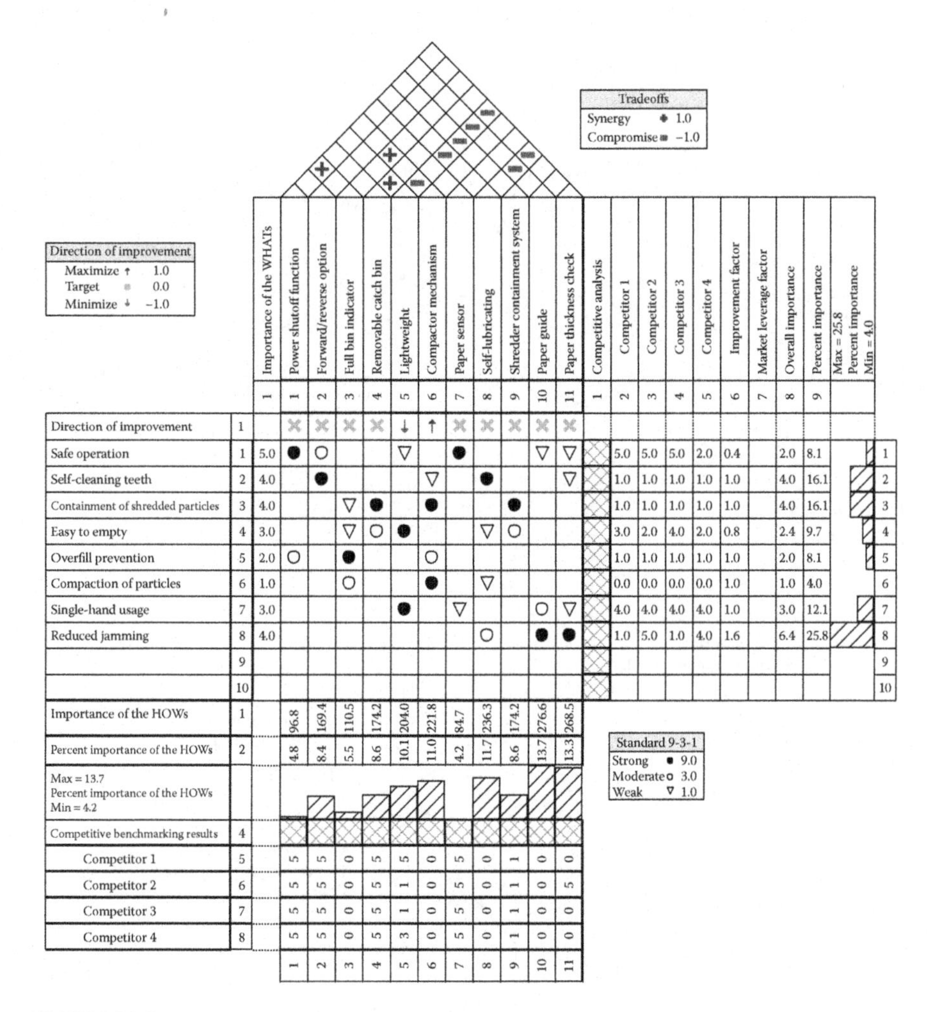

FIGURE 26.6
Complete HOQ.

functional requirements and to add in delighters for the customers as well. Using the HOQ, the team determined from the competitor comparison that there were existing ways in industry to help keep a shredder from jamming, safety features, and methods for keeping the bin from overflowing. All of the designs include "jam-free" technology, which is a sensor in the motor that causes the motor to reverse automatically when it is being overstressed. Safety features include a sensor that detects hands near the shredder opening and stops the motor automatically. Finally, we included a sensor in the bottom of the shredder motor that determines how full the bin is and stops the motor when the bin is full.

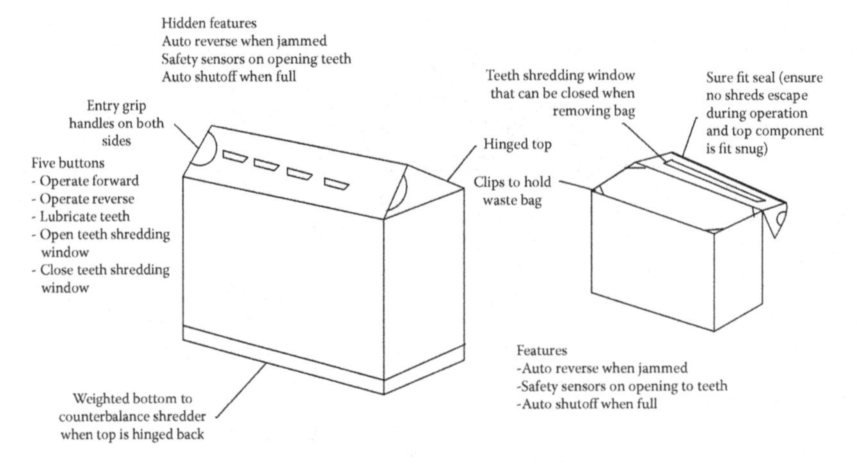

FIGURE 26.7
Concept 1.

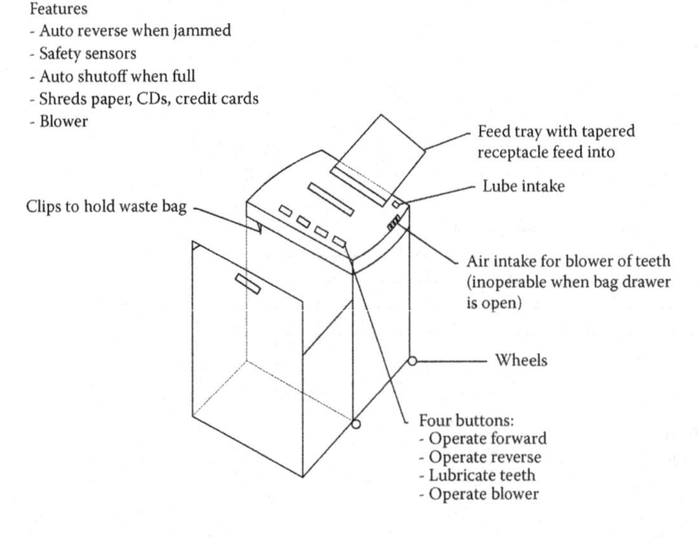

FIGURE 26.8
Concept 2.

Concept 1 (Figure 26.7) uses plastic bags to empty, corresponding to "easy to empty," a top that folds down instead of having to be taken off, and a cover over the teeth to compartmentalize any paper shreds caught in the teeth, which should decrease any mess.

Concept 2 (Figure 26.8) also uses plastic bags in the bin, a tapered paper feed opening to prevent the user from overloading the paper, lubrication to keep the teeth clean, and a blower to knock off paper shreds from within the teeth.

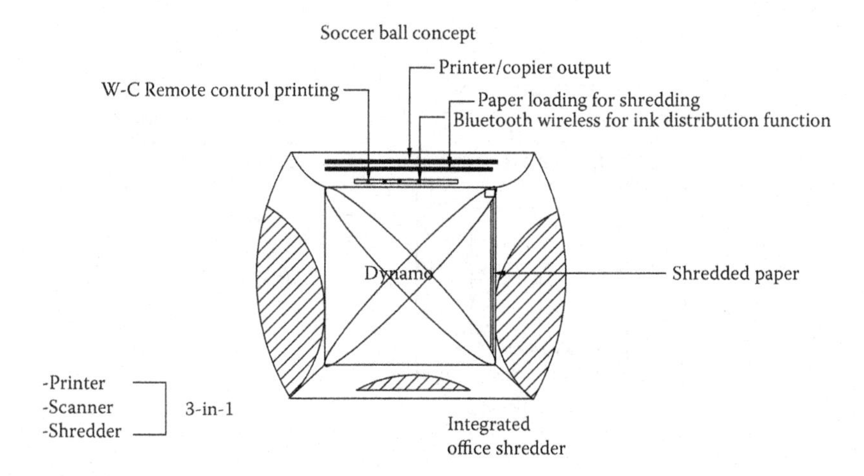

FIGURE 26.9
Concept 3.

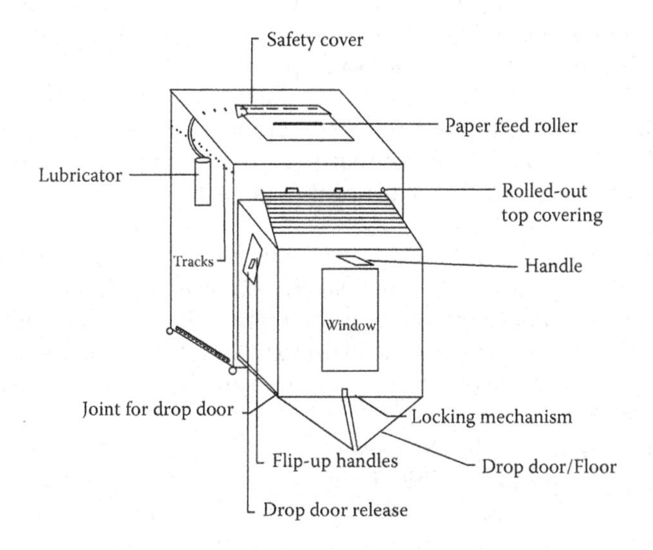

FIGURE 26.10
Concept 4.

Concept 3 (Figure 26.9) is focused on customer delighters, aspects they were not expecting from the paper shredder or did not ask for but would like. As well as providing printing and scanning features along with the shredder, the shedder has an ink distortion feature, which will be a good security measure.

Concept 4 (Figure 26.10) has a lubricator and paper width filter, similar to Concept 2, but also includes a paper roller that will feed the shredder

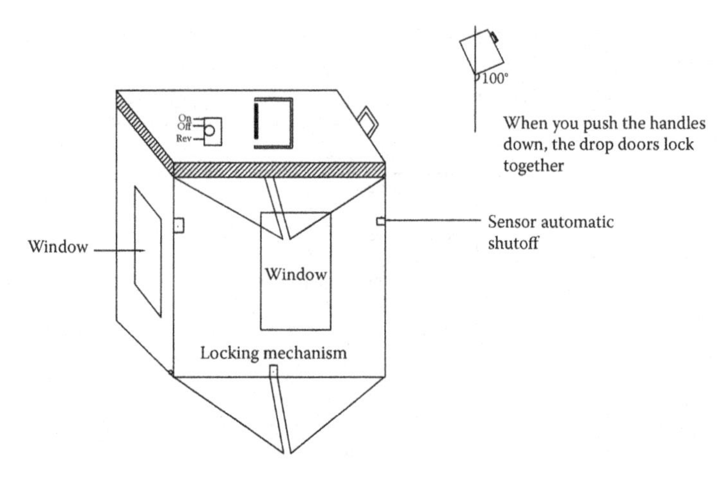

FIGURE 26.11
Concept 5.

automatically. The design includes a rollout cover to trap the shredded particles inside the bin, handles for easy handling, and a drop door in the bottom of the bin to make it easier to empty.

Concept 5 (Figure 26.11) is based on a paper shredder already existing in industry, in which the top is removed to empty the bin. To contain the particles in the bin, there is a cover that is manually put in place by pressing down on the handles on the sides of the bin. The bin will also be emptied from the bottom, as in Concept 4.

Concept 6 (Figure 26.12) adds a special feature, a clean function, that is not available in industry. The clean function is a deliberate process by the shredder to prepare the bin and the teeth to be emptied with minimal mess. The clean function runs the teeth for a couple of seconds, vibrates the entire shredder to settle the shreds, removes any shreds from the teeth, and unlatches the bin.

Concept 7 (Figure 26.13) is focused on providing the customer requirement of compaction of particles. This design includes a track for the compaction mechanism and a handle on the top of the shredder. The handle will be erected from its stored position, and then a plunging action will compact the particles in the bin.

Pugh Concept Selection Matrix

The team evaluated all seven concepts using the Pugh concept selection matrix to determine how the designs rank compared with the datum. In the Pugh concept selection matrix in Figure 26.14, the requirements for safety, overfill protection, and jamming are an S compared with the datum because we were using the datum technology in our designs.

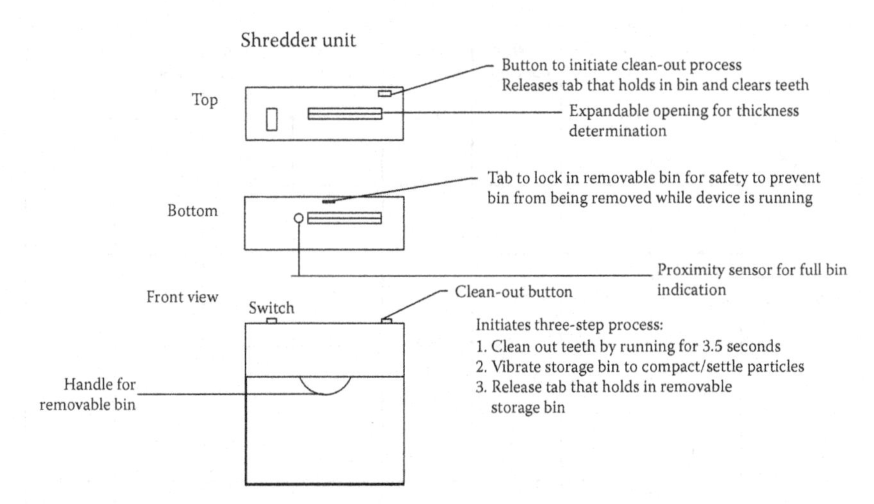

FIGURE 26.12
Concept 6.

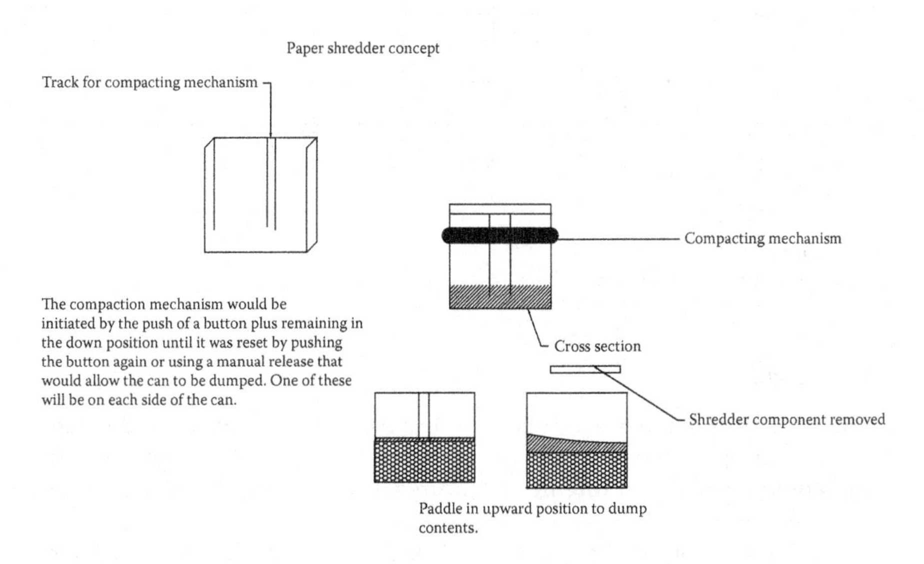

FIGURE 26.13
Concept 7.

Using the Pugh's concept selection matrix, the team was able to determine which concepts were the strongest compared with the customer requirements; Concepts 1 and 4. We were also able to recognize where our concepts were better than the datum, shown as +; therefore, the team decided to add those features to the final design as appropriate.

Criteria	Design concept 1	Design concept 2	Design concept 3	Design concept 4	Design concept 5	Design concept 6	Design concept 7	Final
Safe operation	S	S	S	S	S	S	S	S
Self-cleaning teeth	+	+	–	+	S	+	–	+
Containment of shredded particles	+	S	S	+	+	S	+	+
Easy to empty	+	S	–	+	+	S	S	+
Overfill prevention	S	S	S	S	S	S	S	S
Compaction of particles	S	S	S	S	S	S	+	+
Single-hand usage	S	S	S	–	S	S	S	S
Less jamming	–	S	–	+	–	S	–	S
Sum of (+)	3	1	0	4	2	1	2	4
Sum of (–)	1	0	3	1	5	0	2	0
Sum of (S)	4	7	5	3	1	7	4	4

FIGURE 26.14
Pugh's Concept Selection Matrix.

Final Design

In the final design, the team decided to leave out the automatic feed mechanism in Concept 4 and the ink distortion mechanism in Concept 3, but the team would consider adding these features to special offerings outside our generic shredder.

Our final concept is shown in Figure 26.15. The team included the design in the Pugh concept selection matrix to show that we were able to remove all weaknesses as compared with the datum and capitalize on the strengths of our designs. The final design includes a rollout top and teeth door to compartmentalize the shreds and contain them within the shredder prior to emptying. We have a compaction mechanism, vibrator, and blower to assist in the cleaning and a clean function that does everything the concept provided but adds lubrication to the teeth as well. The team did recognize a limitation in the HOQ: the customers wanted extra functionality to the paper shredder but wanted to keep the shredder lightweight so that it would be

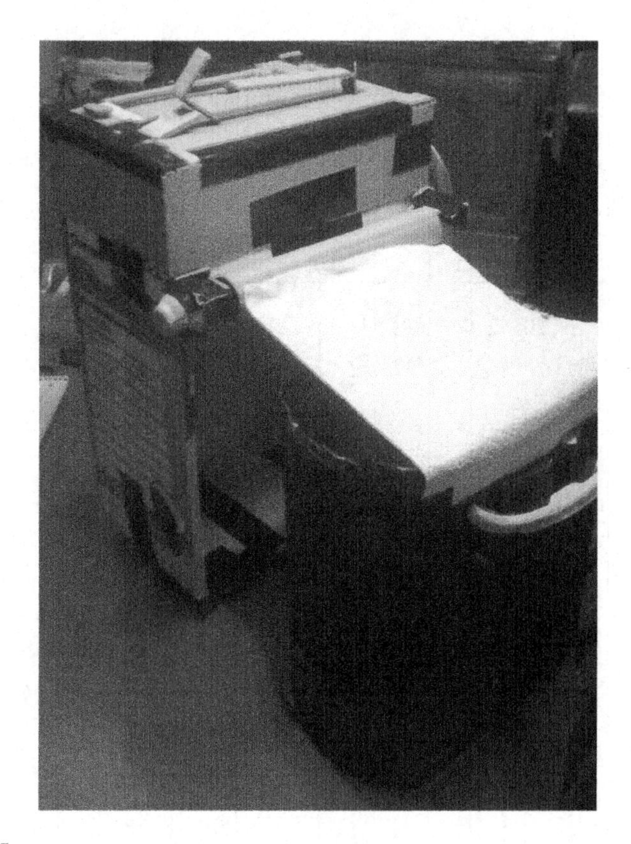

FIGURE 26.15
Final design.

easy to empty. The team decided that a paper shredder on wheels would be easy to move, and leaving the majority of the weight off the bin when the shredder was being emptied would satisfy the customer requirement. A bin that empties from the bottom should take less effort to empty than a design that requires dumping out of the top.

Design Failure Modes and Effects Analysis (DFMEA)

The team used DFMEA to define and qualitatively rank the failure modes of the system and subsystem for the shredder, as shown in Figure 26.16. DFMEA is a standard proactive systematic method for identifying, analyzing, prioritizing, and documenting potential causes of failures. DFMEA would continue to be applied down to subassemblies and components during an industry product development. The team ranked failures that would prevent the system from functioning with a severity of 8; only failures that would result in physical harm to the user were ranked as severity 9. The team would use the SME to help reduce the impact of the failure modes

Potential failure mode and analysis sheet

Potential error mode	Potential effect of failure	Sev	Potential(s) cause mechanism failure	Occ	Current process control	Detect	RPN	Recommended action	Responsibility	Action taken	Action result Sev	Action result Occ	Action result Detection	Action result RPN
1. Paper jam 2. Paper not pulling through	Motor not running	8	1. Motor failure	7	1. Visual inspection 2. Procedure	1	56	Use motors from vendors with proven track record	1. Mechanical systems team 2. Procurement	1. Vendor qualification 2. Trade study to compare motors	4	3	1	12
Shredder blade misalignment	1. Paper not shredded properly	8	1. Blade worn out 2. Blades are misaligned	3	Routine maintenance (procedure) visual inspection	2	48	1. Use parts with track record of performance	Design team	1. Check for expiry date 2. Follow outline procedure	5	1	1	5
1. Shredder doesn't operate 2. Shows full when bin is not full	1. Fault with PCB if no light or displays illuminate 2. Power supply	8	1. Circuit damage 2. Sensor blocked by gum etc.	3	Visual inspection	1	24	1. Design to use on PCB with track record of performance and high reliability.	Systems control team to address issue with trade study.	Trade study	5	3	1	15
1. Disposal paper cabinet not opening	Filled bin cannot be emptied	7	Stuck cabinet unlock lever	6	Routine maintenance (procedure)	1	42	Routine maintenance (procedure)	After market support team	Ensure lock/unlock mechanism is checked periodically	5	2	1	10
1. Fails to roll up cover 2. Lack of power to bring in bin	Won't roll up the cover May not release cover	6	1. Motor failure	4	Routine maintenance (procedure)	1	24	Use motors from vendors with proven track record	1. Mechanical systems team 2. Procurement	1. Durable motors purchased	6	3	1	18
1. Detach or tear 2. Attachment material becomes less secure	Unable to secure shreds from coming out of the top of the bin	5	Adhesive material wear	4	Adhesive replacement	2	40	Acquire proven materials	Design team and procurement	Durable materials purchased and tested	5	3	1	15
1. Unable to compact 2. Unable to be properly stored	1. Unable to open bin	7	Compacting arm bent	5	Visual inspection	1	35	Use higher quality materials	Design team and procurement	Quality materials determined through testing	7	3	1	21
1. Unable to remove excess shreds from teeth 2. Blows shreds out of shredder mouth	Shreds in teeth or coming out of shredder	4	Lack of blower strength or misalignment of blower	3	Visual inspection	3	36	Use motors from vendors with proven track record	1. Mechanical systems team 2. Procurement	Durable motors purchased and testing of alignment	4	2	1	8
1. Unable to be closed 2. Unable to be opened	Bin unable to be placed back under shredder	6	Locking mechanism failure	3	Visual inspection	1	18	Use higher quality materials; test durability	Design team and procurement	Durable materials purchased and tested	6	2	1	12
Unable to be opened	Overheating of shredder; unable to shred due to jams	8	Locking mechanism failure	3	Visual inspection	1	24	Use higher quality materials; test durability	Design team and procurement	Durable materials purchased and tested	8	2	1	16
1. Unable to detect hands near shredder opening	Hand caught inshredder	9	Electrical malfunction or improper calibration	2	Functional testing	2	36	1. Use parts with track record of performance	Design team	Quality materials determined through testing	9	1	1	9

FIGURE 26.16
DFMEA.

identified to assist in testing the products and/or materials to ensure that the failure occurrence and detection decreased.

Optimization

Robustness and Tunability

It is imperative that the ideal function of the product is known and exploited to provide the desired function. Essentially, before testing for optimal performance, there needs to be an agreed-on ideal function on how the product is supposed to perform.

Once the ideal function is known, then the noise factors can be explored. A tool to help identify noise and depict how it is transferred across interface boundaries from the system down to the component level is a system noise map. Additionally, a noise diagram can be used to demonstrate external noises, unit-to-unit noises, and deterioration.

After the various noise factors have been identified, a way to show the effect of noise and control factors on the product is with a P-diagram. The P-diagram illustrates the input/output relationship of parameters that affect performance. This tool shows the quantified result when the product is subject to determined control factors while exposed to noise parameters. When noises have been identified and defined and their impact is known, the product designers can begin the optimization process by exposure to noise and incorporation of a more robust design.

The robustness of the paper shredder refers to the shredder's capability for short-term performance (C_p indices). The tunability of the paper shredder refers to the shredder's capability for long-term performance (C_{pk}) indices. C_{pk} is the measure of capability after the mean has shifted off target by k standard deviations. For both robustness and tunability, these are challenged and measured in the presence of noise factors. In reference to the paper shredder, noise factors represent all forms of incoming variation and include parameters that are not controlled. The noise factors should have a strong effect on the performance, as they add a variety of elements to the shredder beyond its ideal setting and conditions. Examples of noise factors for the paper shredder would be items other than paper, such as wet paper, staples, paper clips, and environmental conditions. These noise factors could affect the paper shredder by jamming it or causing various faults or errors: any response beyond the scope of the intended outcome, or any degradation of the critical functional response (CFR). For this product, the CFR was reduced mess from shredder.

To identify the paper shredder's robustness, several tests would be conducted, and the metric used to measure this CFR would be the amount of visible debris, as this product does not look at the best way to shred paper, but a better way to contain shredded particles and reduce the likelihood of a mess resulting from any operation of the shredder performance cycle, that is, putting paper in to be shredded all the way until the shredder is emptied. Designed noise experiments would be conducted to measure the significant impact of each noise factor on the shredder's intended performance.

A test that could be conducted to determine impacts on the CFR would be attempting to shred and empty the shredder in an office that has a fan or central air conditioning unit. The moving air is outside normal conditions and cannot necessarily be controlled. The results from the test would show where, in the entire process of operating a shredder, a mess of debris occurs. After several noise factors have been examined and tested, that information can be applied to the design of the product to ensure ideal performance in the presence of noise. From the example noise test, a point where the mess of debris could have been identified is when the bin is being brought to the trash can and the fan blows exposed shredding particles out of the bin. This would allow the design engineers to identify flaws in the initial design or add features to improve the design to further reduce the likelihood of that mess occurring.

These tests would generate data that could be used to calculate C_p and C_{pk}. The standard deviations away from the target value or target mess size would show how much variation is in the process (C_p) and then whether the process is centered over the target mess size (C_{pk}). The C_p and C_{pk} values are calculated using Equations 26.1 and 26.2.

$$C_p = \frac{USL - LSL}{6\sigma} \tag{26.1}$$

$$C_{pk} = \min\left[\frac{USL - \bar{y}}{3\sigma}, \frac{\bar{y} - LSL}{3\sigma}\right] \tag{26.2}$$

Once the C_p and C_{pk} are known, the product can be optimized by increasing its robustness and reducing the variation from noise while ensuring the performance is meeting the target values.

System Additive Model

During critical parameter management, the data that is collected needs a mechanism to validate the robustness and capability of the critical parameter relationships. A way of doing just that is the system additive model. Mathematically, the additive model is shown in Equation 26.3.

$$\begin{aligned} S/N_{opt} &= S/N_{avg.} + (S/N_{A\,opt.} - S/N_{avg.}) + (S/N_{B\,opt.} - S/N_{avg.}) \\ &\quad + \cdots + (S/N_{n\,opt.} - S/N_{avg.}) \end{aligned} \tag{26.3}$$

The essential element in this model is the signal-to-noise (S/N) ratio. This ratio is found during the noise testing and is a metric that represents the gains in robustness and is applied to the additive model to validate the results. In the additive model, control factors are represented by A and B and the resulting. This validates the data, as it quantifies system improvement as controls were added to reduce variation in the presence of noise. For the paper shredder, an example of a control factor would be a more secure bin cover that has a rubber seal to create an airtight compartment.

Verify

Customer Feedback

The team conducted a face-to-face survey to get customer feedback on our design to verify that the final design meets the customers' demands and to determine whether there were areas of the design that could be improved on. The survey results are shown in Figure 26.17.

Important feedback for the team was that all the customers surveyed liked being able to empty the shredder bin from the bottom and thought that this would make it easier to empty the shredder bin. Also, the clean function was thought by the majority of the customers to be value added and beneficial for keeping the shredder clean. The customers thought that the added features would keep the process of emptying the shredder less messy, but were concerned about the price for the added features. For this product, the team assumes that customers might pay more than they originally indicated in the survey for a cleaner paper shredder; otherwise, we might not have the scope to meet the customer demands. The customers also had concerns over the durability of the compactor mechanism, but this could be due to the prototype consisting of flimsy cardboard, and that is why we included the compactor mechanism in our failure modes and effects (FMEA) analysis. Since this was a new concept, our customers stated that we would need to provide some education on how to use the product, but the team believes that the product would meet the customers' needs once they understand how to use it.

Robustness Evaluation

The final robustness evaluation the team would conduct is on the final product design to include all the changes made during initial robustness and tunability testing and DFMEA factors. The system robustness would be tested in nominal conditions and in conditions of stress to test how the entire system performs prior to final production and shipping to stores. The team

Customer 1 (Female, 25–30)		Customer 2 (Female, +50)		Customer 3 (Male, +50)		Customer 4 (Male, 25–30)	
Likes	Concerns	Likes	Concerns	Likes	Concerns	Likes	Concerns
Features (clean, vibrate, blower, compaction)	Concern with using velcro with top cover ... how long would it last	Concerned about the ability of blower to clean teeth (we should test it)	Have concerns with additional price tag	Likes that our shredder has mechanism to help prevent jamming	Concerns with the compactor	Concern that the motors required for the blower, vibration, and cover would add too much cost	Concerns about how durable the compactor is over time (we should test it)
Bottom empty feature on paper bin	Concern with durability of compactor	Thinks the vibration function would be very beneficial	Design would require consumer education to successfully use new features	Likes the cleaning feature but would need to educate the customer how often to use it	Have concerns with additional price tag	Likes clean Function	Make cover mechanism likes a tape measure (loads like a spring instead of using a motor)
Likes the paper amount filter	Vibration could scratch wood floor (maybe)	Loves the "clean" function (and the lubrication that occurs ... thinks it will help a lot)	Would not pay more than $5-6 for clean function (specifically clean function)	Likes the trapdoor	Design would require consumer education to successfully use new features	Likes trapdoor	
Likes trapdoor		Thinks the cover will help reduce static electricity		Should make it more streamline (the outside) to fit in small places		Likes top cover (of bin)	
Likes top cover (of bin)		Likes emptying from the bottom					

FIGURE 26.17
Customer survey on final design.

would want to test how long the motors last with constant use, how long our cover lasts before the Velcro material and snaps fail, how many times we can close the bottom of the bin before the locking mechanism fails, and other features to determine failure points. The team would also want to test and evaluate system reliability based on mean time to failure. We would want to verify that product design meets customer requirements based on our CFR specifications. Finally, the team would want to verify that the production is ready and capable of producing the design with a very low percentage of defects. We would want to run our production line for a test batch of products and then assess the reliability of our production. All these evaluation steps and testing are important prior to full operation, because it is cheaper to fix problems prior to full production than to recall items after production.

Conclusion

Using DFSS tools, the team was able to develop a robust product design that pushes the current industry limits for paper shredder design. The customers would need some education on how to use new features on the paper shredder, but we think that they would like to see these features on store shelves in the future. Currently, the paper shredding industry has focused primarily on the safety and function of the shredder itself and has not considered how to make the paper-shredding process hassle free when emptying. Our design would give a company a start in the direction of entering this new market. During production, we would have to consider the costs of the added features, but we would assume that there is a market for a cleaner shredding process. The data shows that the majority of customers would only pay $20.00 extra for a clean paper shredder. Our design could be scaled to meet customer demands that are more price sensitive and also scaled for the more deluxe designs on the market. DFSS tools focus developers on considering more than just design, and as a result, the customer and the company get a better product.

27

Design for Six Sigma Case Study: Universal iPhone Dock

Hanan Altabbakh, Charlie Barclay, Amita Ghanekar, and Elizabeth Cudney

Design for Six Sigma (DFSS) is a set of best practices and tools which, when integrated into a product or service development process, increases an organization's ability to meet customer requirements in a timely and cost-effective manner. As the name suggests, DFSS is about design. DFSS saves costs in part because the cost of making a change increases exponentially during the life cycle of a development project. DFSS seeks to avoid manufacturing/service process problems at the outset (i.e., fire prevention) by using advanced voice of the customer (VOC) techniques and proper systems engineering techniques. When combined, these methods obtain the proper needs of the customer, and derive engineering system parameter requirements that increase product and service effectiveness in the eyes of the customer and other people.

Project Description

A docking station provides a simplified way of plugging-in an electronic device such as a laptop computer to common peripherals. Many different docking stations are currently on the market, designed to fit many different needs. Owners of Apple products and, in particular, iPhone owners, have a multitude of docking stations to choose from. The most common style of docking station allows users to charge their phone and listen to music that is stored on their phone through speakers embedded in the docking station.

Docking stations are common and widely available. What is not common or widely available, however, is a docking station that will allow iPhone users who choose to protect their phone with an OtterBox case, the opportunity to dock their phone without first removing the case. The

OtterBox Defender Series cases are among the most durable, but also the most bulky cases available. These cases are designed to keep phones safe when exposed to impacts or shocks, and they do this at the expense of compactness. These cases make "slim" smartphones much bulkier and less compatible with phone accessories. Of interest here is the fact that these cases make iPhones incompatible with currently available docking stations. The current options on the market for iPhone users with an OtterBox require "slight tweaking with a Dremel tool." Although some "do-it-yourselfers" would have no qualms about attacking a recently purchased electronic device with a rotary tool, many might not be so willing. Many people want to purchase a product that works right, directly out of the box. This project seeks to offer a solution to iOs (iPhone with an OtterBox users) who would like to be able to charge their phone and listen to music at the same time.

Project Goals

As discussed in the previous section, there currently exists no "out of the box" solution for iOs users who wish to purchase a docking station. This leaves a gap in the market that needs to be filled. The goal of this project is to design a product to fill that gap the through utilization of the DFSS methodology by providing iOs users the opportunity to enjoy the benefits of a docking station without first having to remove their phone from its case or without having to modify a current docking station with a Dremel tool, thereby foregoing any warranty offered to them by the manufacturer. The team believes that by applying the DFSS principles, we will be able to be first to market, with a product that will work correctly, without modification, right out of the box, and provide the ability to charge and listen to music simultaneously.

Requirements and Expectations

The requirements and expectations for this project consist of the following:

- Clearly define what customers want/need in a docking station
- Create several design alternatives that fulfill the wants/needs of the customer
- Use an engineering approach to compare design alternatives with each other and a competitors' design considered best-in-class
- Modify design alternatives to arrive at one final design that will satisfy customer requirements and allow for use with OtterBox cases
- Perform a failure modes and effects analysis (FMEA) on final design
- Prototype final design

Project Boundaries

In order to keep the project manageable in the allotted time frame, the team focused the scope of this project on only a few versions of iPhones. The team considered iPhone versions 3 and 4 (including all variations, i.e., G, S, etc.) when encased in an OtterBox Defender Series protective case. These cases are considered to be the most bulky cases on the market, and often impinge on the phone's compatibility with other accessories. If, along the way, there seems to be a simple solution to allow for expanding the scope to other phone versions, then the team may consider that at the appropriate time.

Project Management

In order to facilitate the completion of tasks and project milestones, a Gantt chart was created. This tool was used throughout the course of the project. The Gantt chart was the schedule by which the team measured their project completion rate throughout the completion of the project. The complete Gantt chart can be seen in Figure 27.1. As shown in the chart, a large portion of time was dedicated to concept generation and optimization. This was done to allow the design team sufficient time to generate, evaluate, and compare concepts. This large time block also allowed for multiple iterations on design solutions.

Voice of the Customer

VOC is a process that is defined as "capturing the customer expectations." It is mainly a market research technique that produces a detailed set of customer wants and needs, organized into a hierarchical structure, and then prioritized in terms of relative importance and satisfaction with current alternatives. In this project, the VOC was established by carefully constructing a survey that was sent out to a wide variety of people covering many demographics. These results were then categorized using statistical analysis and visually through an affinity diagram. The results of these tools were then used to construct a house of quality (HOQ). The HOQ was then used by the design team to create preliminary designs.

Customer Survey

In order to measure the VOC, a survey was constructed and sent out to the public. A questionnaire with 11 items including 2 demographic questions was used to collect the data. The goal of the survey was to determine the

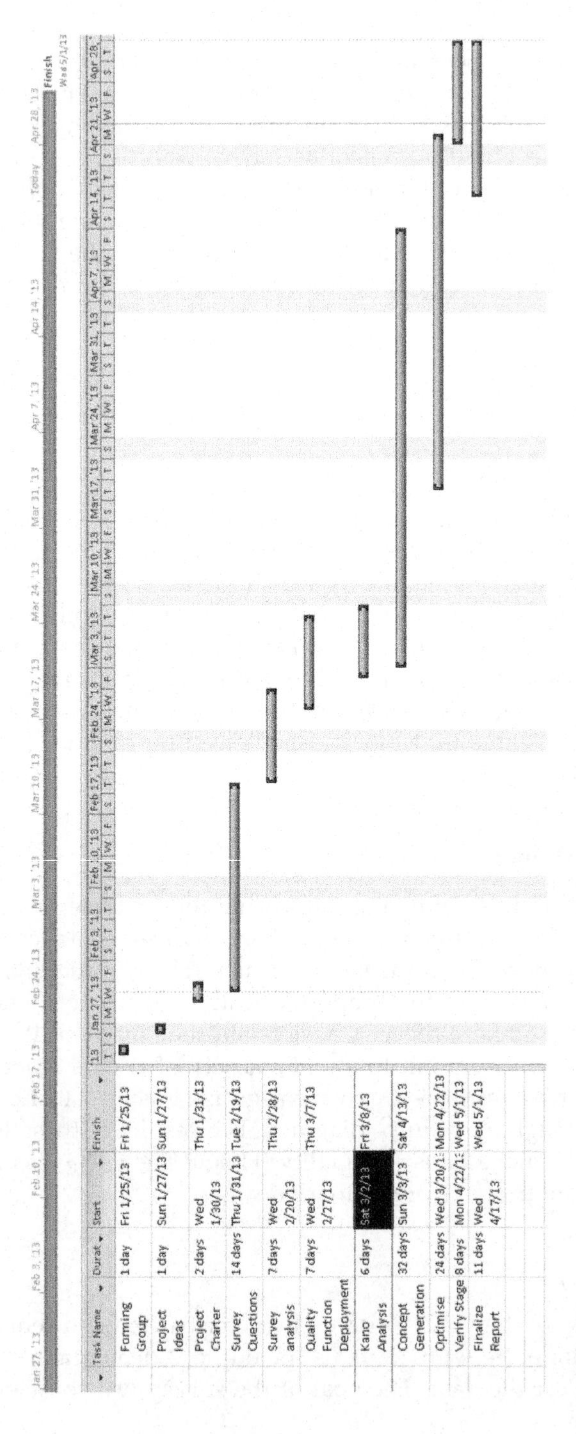

FIGURE 27.1
Gantt chart.

customers' requirements, complaints, suggestions, and desires of a docking station. One question was asked to determine if the respondents owned a smartphone. One question with subcategories was asked about the type of case they currently use or have previously used. Two questions were asked about whether or not they own a docking station and the location where they usually keep it. Five questions were asked to determine customers' expectations of a newly designed docking station.

Survey Results

A total of 179 questionnaires were returned. However, not all of the surveys were filled out entirely. Of those responding to the survey, 73% were male, 26% were female, and 1% preferred not to answer. The age demographics of the respondents were as follows: 44% were between the ages of 18 and 25, 40% were between the ages of 26 and 35, 11% were between the ages of 36 and 50, and 5% were between the ages of 51 and 70. Additionally, 80% of the respondents have used a protective case for their smartphone where the other 20% did not. Of the 20% that did not own a protective case, 21% responded that it would be too bulky to have one. Of the 80% that did own a protective case, 22% had the OtterBox type of protective case. As shown in Figure 27.2, another 16% of the respondents claimed that the case they were using prevented them from using certain phone accessories. Further, 70% of the respondents currently own a docking station, previously owned, or would consider purchasing one, while the other 30% would not consider buying a docking station. When asked where they would like to use a docking station, 12% of the respondents said they would use their docking station in the office, 33% in the bedroom, 12% in the kitchen, and 15% would use it in other places such as while camping, as shown in Figure 27.3. When asked how important aesthetics were in their decision to purchase a

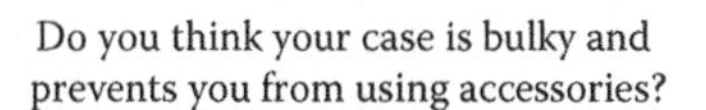

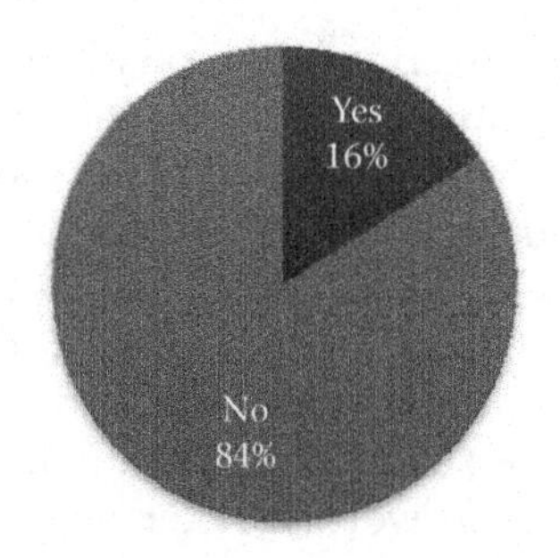

FIGURE 27.2
Survey results relating to bulky cases preventing accessory use.

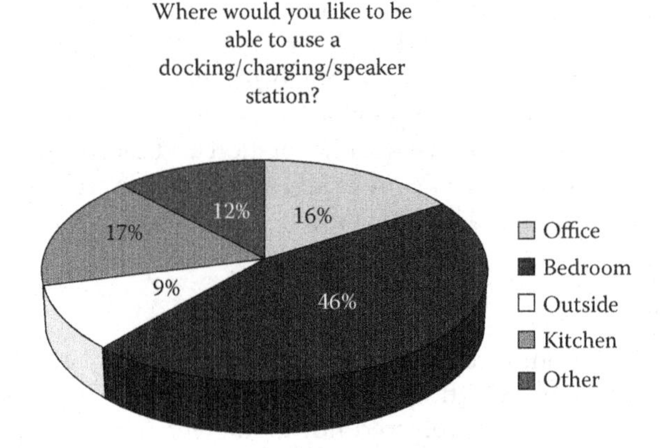

FIGURE 27.3
Survey results relating to where the docking station would be used.

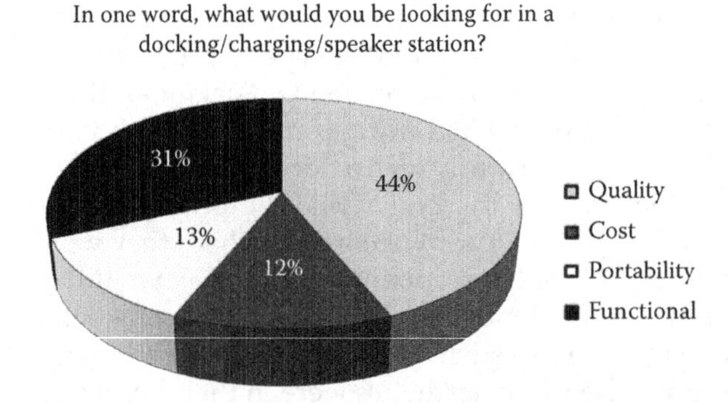

FIGURE 27.4
Survey results relating to the most important feature in a docking station.

docking station, 12% of the respondents considered aesthetics very important, 39% considered it important, 38% were neutral, and only 5% thought it was unimportant. As shown in Figure 27.4, the quality of the docking station was the most desired feature with a weight of 44%, functionality 31%, cost 13%, and portability 13%. Additionally, 48% of the respondents were willing to pay a price of between \$25 and \$50 for a new docking station, 27% between \$0 and \$25, 19% between \$50 and \$75, 5% between \$75 and \$100, and only 1% would pay more than \$100. Finally, the top three desired features in a docking station were speaker quality, cost, and portability. Table 27.1 gives a breakdown of the overall importance of the following features:

TABLE 27.1

Which Three of the Following Features Do You Consider to Be the Most Important?

Speaker quality	19
Cost	16
Portability	13
Wireless capability	9
Size	9
Remote control	6
Cover for speakers	5
Multi phone compatibility	5
Shape	4
Weight	4
AM/FM	3
Worldwide elec. compatibility	3
Color	2
Large display	2

Affinity Diagram

After reviewing the results of the survey, it was time to analyze the responses of the customers and evaluate their needs and requirements. It was essential to group the data into key issues under labels that reveal customers' needs using an affinity diagram as depicted in Figure 27.5. These key issues are then incorporated to determine the nine concepts of the newly designed docking station.

Quality Function Deployment (QFD)

Based on the results of the customer surveys and the affinity diagram, a HOQ was constructed. The customer requirements were extrapolated directly from the statistical analysis performed on the survey results. These requirements were the characteristics that were most commonly considered to be the most important. The weighting for each customer requirement was also drawn directly from the survey results. Many of the survey questions were formulated in such a way as to easily extract statistical averages associated with each customer requirement. The functional requirements were developed by the design team. These were characteristics that were deemed necessary to be able to produce the product as demanded by the customer. Target or limit values were selected by the design team in such a way as to deliver a product of exceptional quality, while at the same time deliver a product that was cost-effective. A competitive analysis was performed on three products currently produced by competitors. Each team member evaluated the three competing products individually and

Cost	Safety	Advantages	Portability	Compatibility
Price	Durability (withstand)	Musical alarm	Portable (camping)	Utility
Value	Fit (secured phone)	Multipurpose	Wireless	Compatibility, with other phones
			Hands-free	USB-compatible

Quality	Sound output	Usage	Aesthetics
Quality Reliability	Music Bass	Practicability Functionality	Sleek Stylish
Sound quality	Sound	Ease	Size
Speakers quality	Loud	User-friendly	Dark color
	Sound, hear it throughout apartment while in one place	Convenience	Feng shui
		Usability, easy to use	Small
			Excellent design inconspicuous small compact lightweight

FIGURE 27.5
Affinity diagram constructed from the survey results.

the ratings were then averaged to give the results shown. The HOQ can be seen in Figure 27.6.

Kano Analysis

The Kano Model is a theory of product development and customer satisfaction developed in the 1980s by Professor Noriaki Kano, which classifies customer preferences into five categories. The Kano Model defines three types of quality requirements:

- One-dimensional quality
- Expected quality
- Exciting quality

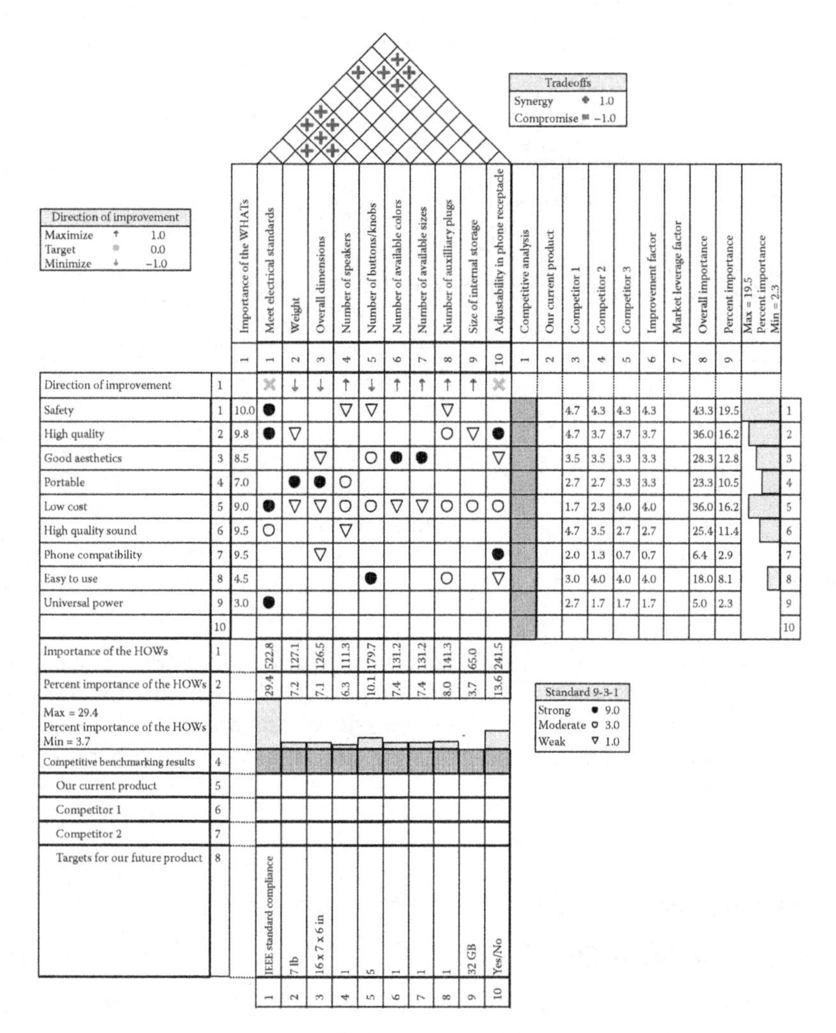

FIGURE 27.6
House of quality that was constructed based on the survey results.

The Kano Model also defines how the achievement of these requirements affects customer satisfaction. Figure 27.7 represents a Kano Model, which divides the requirements into basic needs, performance needs, and delighters. Doing a Kano Model addresses the three types of requirements:

- Satisfying basic needs: Allows a company to get into the market
- Satisfying performance needs: Allows a company to remain in the market
- Satisfying excitement needs: Allows a company to excel, to be world class

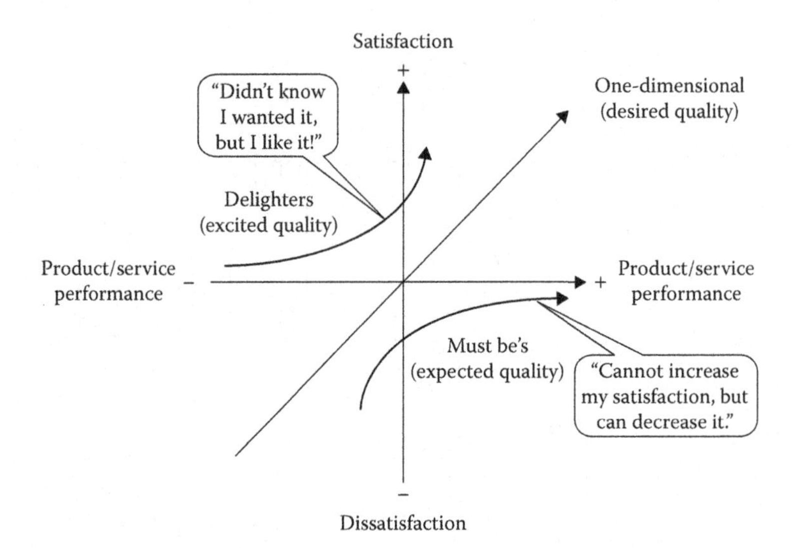

FIGURE 27.7
Kano Model.

The Kano Model of Customer (Consumer) Satisfaction (KMCCS) classifies product attributes based on how they are perceived by customers and their effect on customer satisfaction. These classifications are useful for guiding design decisions in that they indicate when good is good enough, and when more is better.

One-Dimensional Quality

The one-dimensional qualities are also known as the *threshold qualities*. These are the expected attributes or "musts" of a product, and do not provide an opportunity for product differentiation. Expected features or characteristics of a product or service (legible forms, correctly spelled name, basic functionality) are typically "unspoken." If these needs are not fulfilled, the customer will be extremely dissatisfied. For our product, we have taken into consideration the following one-dimensional qualities: speakers, auxiliary input, and electrical plug. Additionally, the docking station must work correctly and be safe to use. We specifically considered these five as our one-dimensional qualities as they are the basic needs of a docking station.

Expected Quality

The expected qualities are also known as the *performance attributes*. The price that the customer is willing to pay for a product is closely tied to performance attributes. These are standard, or expected, product characteristics that increase or decrease customer satisfaction by the degree (cost/price, ease of use, speed) at which these characteristics are met. These needs are typically "spoken." As specified, these are the qualities that are neutral in designing the product. The expected qualities of our product are: radio (AM/FM), alarm clock, battery operated, remote control, charging station. These are some of the expected qualities of a docking station, which if unfulfilled, cause dissatisfaction to the customer.

Exciting Quality

Excitement attributes are unspoken and unexpected by customers but can result in high levels of customer satisfaction; however, their absence does not lead to dissatisfaction. Unexpected features or characteristics which impress customers and earn the company "extra credit." These needs also are typically "unspoken." These are the qualities that make customers extremely happy and satisfied. Our docking station will have the following exciting qualities: compatible with other phones (Android as well as iPhones), better aesthetics, portable, and easy to carry anywhere, and universally electrically compatible. These qualities make our docking station better than other docking stations.

Concept Generation

Concept generation was completed using traditional brainstorming techniques. Ideas for design possibilities were discussed among team members and each team member then developed three designs for consideration. A total of nine concepts were generated. The nine concepts, labeled A through I are shown in Figure 27.8.

Pugh Concept Selection Matrix

In order to refine the original nine concepts, a Pugh concept selection matrix was completed. Table 27.2 shows the Pugh concept selection matrix that was completed for this project. The Pugh concept selection matrix allowed for comparison between the nine concepts and the best-in-class product, or datum (in this case a Bose system). The comparison allowed for viewing design weak points and allowed for further refinement of the nine concepts, combining strong features of concepts to eliminate the weak features in other concepts. Using the Pugh concept selection matrix, the original nine concepts were reduced to three. This was done by analyzing the features from the nine concepts that were better than the datum, and using them to replace features on

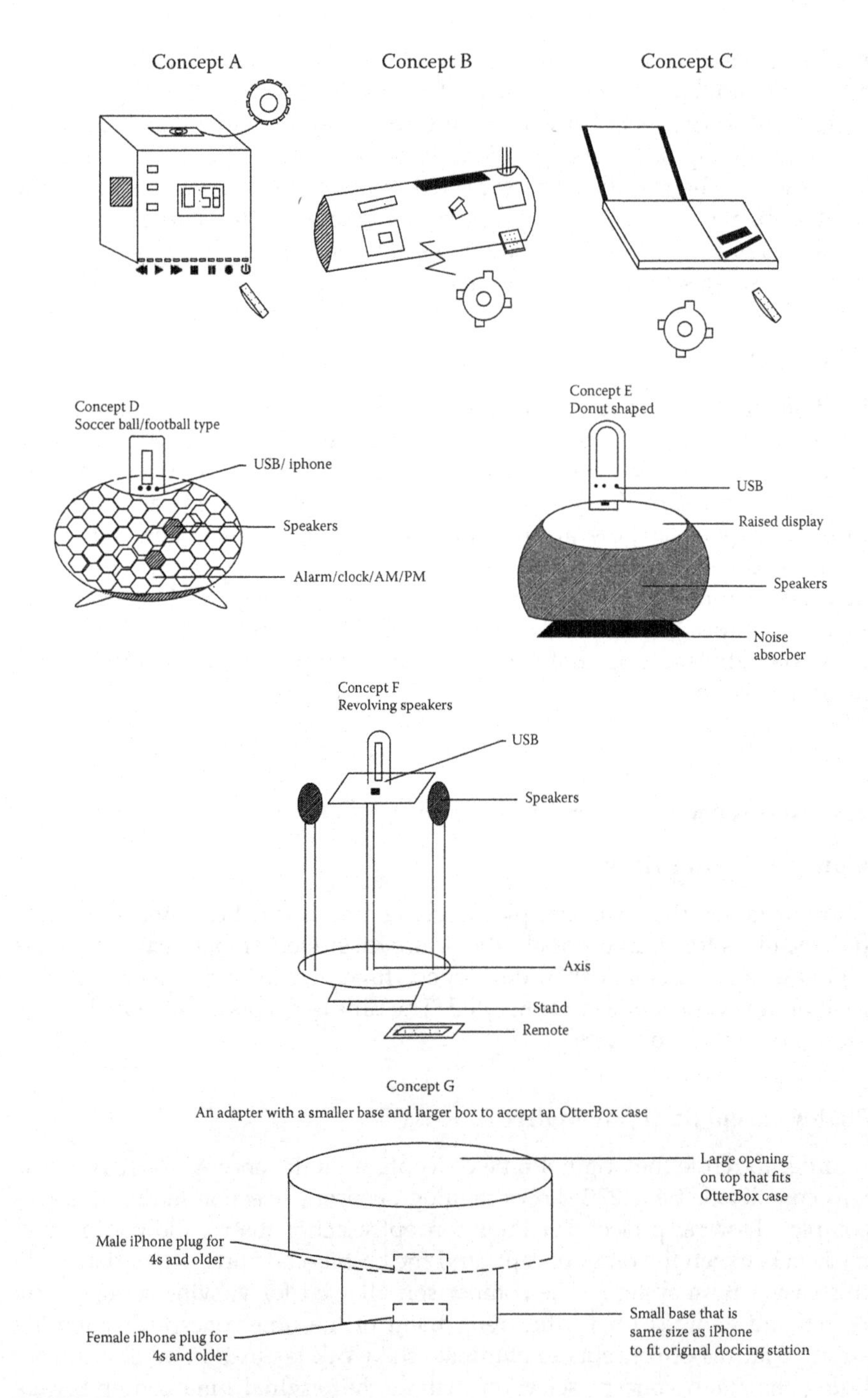

FIGURE 27.8
Original nine docking station concepts.

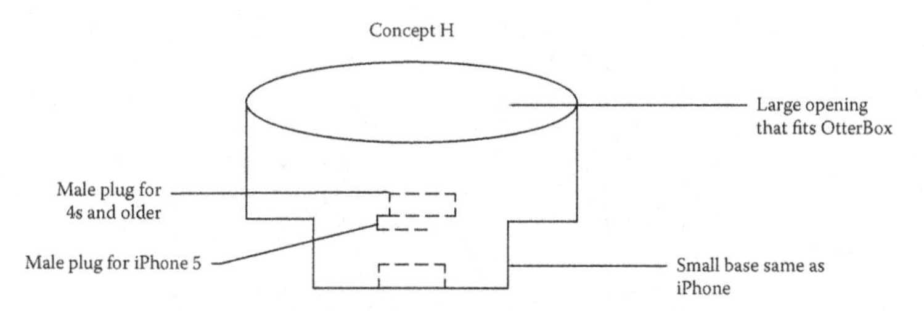

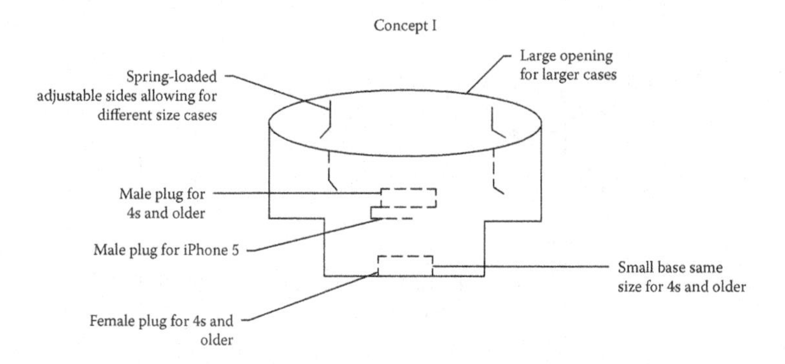

FIGURE 27.8 (CONTINUED)
Original nine docking station concepts.

other concepts that were inferior to the datum. The "Sum of (+)" shows the features that were considered better than the datum. The "Sum of (–)" were the features that were considered inferior to the datum, and the "Sum of (S)" were the features that were considered the same as the datum. The left-hand side of the table shows the analysis that was completed on the nine original concepts. The right-hand side shows the analysis that was completed on the three, refined concepts. The final concept was created by combining the best attributes of concepts J, K, and L, eliminating the weak points in the final design.

Final Design

The final concept was chosen based on the Pugh concept selection matrix. The concept was a combination of features from several of the original concepts. Figure 27.9 shows the final concept. The final design took the form of

TABLE 27.2

Pugh Concept Selection Matrix for Concept Refinement and Selection

Criteria	Concept A	Concept B	Concept C	Concept D	Concept E	Concept F	Concept G	Concept H	Concept I	Concept J	Concept K	Concept L	Datum
Safety	s	s	s	s	s	–	s	s	s	s	s	s	//
High quality	–	–	–	–	–	–	s	s	s	s	–	–	//
Good aesthetics	s	–	+	s	s	s	+	+	+	+	+	+	//
Portable	+	s	+	+	+	–	+	+	+	+	+	+	//
Low cost	+	+	+	+	+	+	+	+	+	+	+	+	//
High quality sound	–	–	–	–	–	–	s	s	s	s	–	–	//
Phone compatibility	+	+	+	–	–	–	–	+	+	+	+	+	//
Easy to use	s	s	+	+	+	+	+	+	+	+	+	+	//
Universal power	+	+	+	–	+	+	+	+	+	+	+	+	//
Sum of (+)	4	3	6	3	4	3	5	6	6	6	6	6	//
Sum of (–)	2	3	2	4	3	5	1	0	0	0	2	2	//
Sum of (S)	3	3	1	2	2	1	3	3	3	3	1	1	//

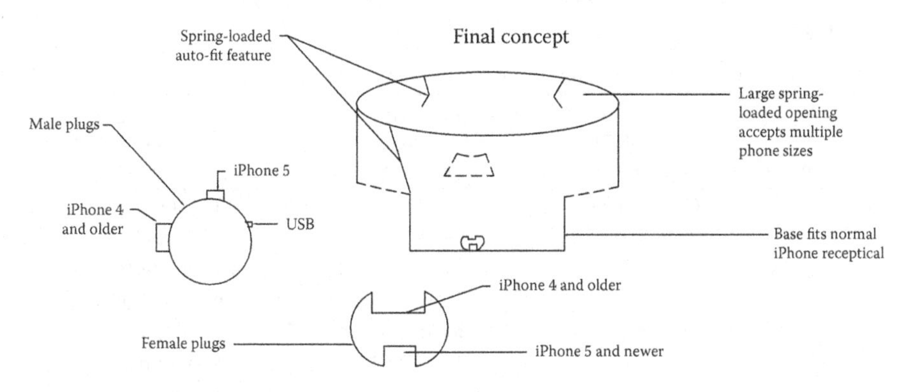

FIGURE 27.9
Final design selected from the Pugh analysis.

an adaptor. This design was chosen based on its ability to fill a wide range of customer requirements. While this device is not stand alone, it does fill a gap in the market. Being able to use this device will allow users to select the specific type of docking station that best suits their needs, and then simply use this adaptor to allow their phone to fit. The device will be cost-effective to manufacture, and could easily be produced in a wide range of colors. This adaptor will allow for using multiple phones in any iPhone docking station.

The top of the adaptor is large enough to accept any iPhone (and other smartphones), even when using an OtterBox. The sides of the top include an auto-fit option, which consists of a spring-loaded mechanism that will open and close as a phone is placed inside. The sides will be tapered so that when a phone is placed into the adaptor, the sides will retract, and will stay snugly seated against the front and back of the phone.

In addition to the auto-fit feature, the adaptor also consists of two separate plug-selection wheels. The top wheel consists of three male plugs; iPhone 5 and newer, iPhone 4 and older, and USB. This allows the user to rotate the wheel to match the type of phone that he or she is using. The second wheel is in the base and consists of two female plugs; iPhone 5 and newer, and iPhone 4 and older. Again, this allows the user to select the right plug based on the type of docking station that he or she owns.

Design Failure Modes and Effects Analysis (DFMEA)

DFMEA is also known as *potential failure modes and effects analysis;* failure modes, effects, and criticality analysis (FMECA). Failure modes and effects analysis (FMEA) is a step-by-step approach for identifying all possible

TABLE 27.3

FMEA Produced for the Selected Concept

Item/Function	Potential Failure Mode	Potential Effect(s) of Failure	SEV	Potential Cause(s)/ Mechanism(s) of Failure
Electrical circuit	Current loss	Unit will not operate	5	Bad plugs
			5	Bad wires
			5	Bad connections
			5	Improper wiring
Auto-fit feature	Improper closing	Auto-fit will not close properly	2	Springs jam
			2	Springs break
			2	Dirt/debris
	Improper opening	Auto-fit will not open properly	3	Springs jam
			3	Springs break
			3	Dirt/debris
Plastic case	Cracking/breaking	Unit unusable	5	Improper usage
			5	Manufacturing defect

failures in a design, a manufacturing or assembly process, or a product or service. "Failure modes" means the ways, or modes, in which something might fail. Failures are any errors or defects, especially ones that affect the customer, and can be potential or actual. "Effects analysis" refers to studying the consequences of those failures. Failures are prioritized according to how serious their consequences are, how frequently they occur, and how easily they can be detected. The purpose of the FMEA is to take actions to eliminate or reduce failures, starting with the highest-priority ones. FMEA also documents current knowledge and actions about the risks of failures, for use in continuous improvement. FMEA is used during design to prevent failures. Later, it is used for control, before and during the ongoing operation of the process. Ideally, FMEA begins during the earliest conceptual stages of design and continues throughout the life of the product or service. Begun in the 1940s by the U.S. military, FMEA was further developed by the aerospace and automotive industries. Several industries maintain formal FMEA standards. The FMEA is provided in Table 27.3.

Final Design Prototype

A prototype was created to illustrate the idea to potential customers. The four images in Figure 27.10 are of the final design prototype. The top left image shows the disassembled prototype. The top piece is the cradle in which the iPhone sits. The bottom piece is the base, which plugs into the

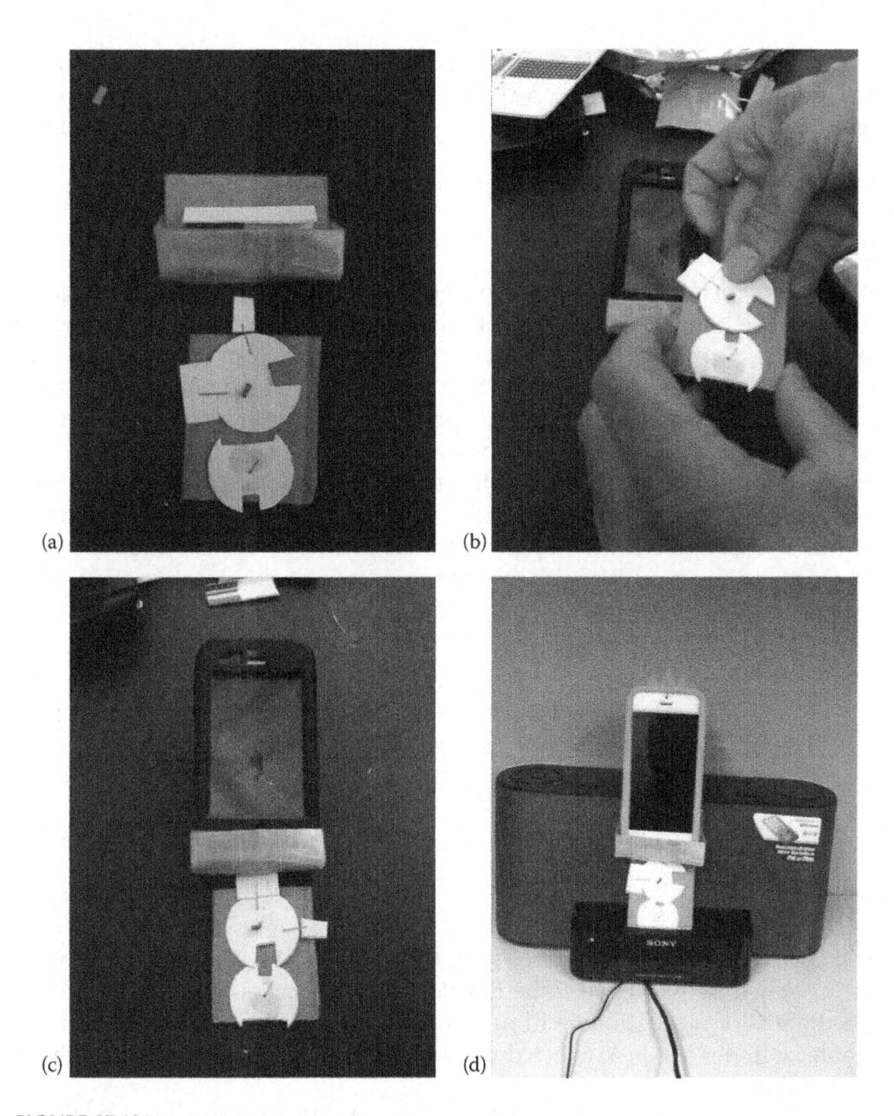

FIGURE 27.10
(a) The prototype; (b) showing the rotating function of the plug selection switch; (c) showing the adapter with an iPhone 3 in an OtterBox; (d) showing the adapter with an iPhone 5.

docking station. The two wheels on the base are the selector switches, which allow the user to plug in different phones. These work by rotating the switch to the required plug, as shown in the top right image. The top selector switch has an iPhone 4 (and older) male plug, an iPhone 5 (and newer) male plug, and a USB female plug. The bottom switch has an iPhone 4 (and older) female plug and an iPhone 5 (and newer) female plug.

The bottom two images show the prototype being used in two separate configurations. The bottom left image shows the prototype configured to accept an iPhone 3GS (with an OtterBox) and plug into an iPhone 4 (or newer) docking station. The image on the bottom right shows the prototype configured to accept an iPhone 5 and plug into an iPhone 4 (or newer) docking station.

Verification

In order to verify the concept, it was deemed important to put the prototype into the hands of the customer. This was done to allow the customer to see how the product would function, and give the customer an opportunity to provide unprompted feedback. Several potential customers were chosen and given an opportunity to see the prototype, given a brief demonstration on the functionality of the prototype, and then asked to write down any comments they may have on the design. Overall, the design was very well received. The customers liked the idea of not having to purchase an entirely new docking station if they upgraded their phone. They also liked the idea of being able to use multiple phones with multiple docking stations. Here are the most common responses after viewing the prototype:

- "The current prototype is too big. It needs to be smaller."
- "The fact that the adapter can be used for different phones and different docking stations is a nice feature."
- "The addition of the USB plug is nice."
- "I would buy this."

Testing

In order to remain in compliance with U.S. laws and regulations, a thorough testing schedule will need to be developed and administered. At least two specific sets of regulations will need to be thoroughly investigated and the adapter will need to be subjected to testing to ensure compliance with these regulations.

The first set of regulations will be from the Institute of Electrical and Electronics Engineers (IEEE). This body establishes electrical standards for products developed, and sold in the United States. The specific regulations relating to this product will need to be established and then a thorough testing schedule developed and executed. This testing will check all of the electrical circuitry.

The second set of testing will ensure that the product meets standards established by the Consumer Product Safety Commission. This set of testing will ensure that the product meets general safety requirements and will not pose a danger to any potential customers.

Conclusion

Overall, the team was very pleased with the results of this project. The team was able to use DFSS tools and methodologies throughout the course of this project to capture the VOC and translate it into a prototype concept.

DFSS tools were used to design and develop a product to fill a gap existing in the market. Currently, if an iPhone user's case is too bulky, then he or she must manually alter a docking station (with a Dremel or other rotary tool) to allow his or her phone to fit. This product would give that market segment another option. The selected design was successfully prototyped and placed in the hands of the customer. The prototype was verified and approved by the potential customers.

28

Design for Six Sigma Case Study: Hospital Bed

Bhanu Partap Singh Kanwar, Adam Twist, Anthony Masias, and Elizabeth Cudney

Designing a Hospital Bed for Improving Stakeholders' Level of Care

The mission of this Design for Six Sigma (DFSS) project is to create and standardize the industry bed by integrating a universal bed that will provide a safe and comfortable environment to the patient and staff members. For the team to be able to provide this service, we are redesigning the current standard hospital bed in order to provide stakeholders with a more efficient, safe, and robust bed.

Design for Six Sigma Overview

In order for the team to develop a viable product, the primary goal is that once it is introduced to the market the hospital bed will have minimal to no defects. Because of this requirement, the team utilized the DFSS concept in order to facilitate the product development process. Simply put, DFSS integrates three major elements for product development: keeping cost low, producing a high-quality product, and moving from concept design to market quickly. DFSS is a data-driven quality strategy that interconnects five phases; therefore, the team is able to utilize its process of define, measure, analyze, design, and verify (DMADV) to develop the ideal bed.

Project Description

The goal of the DFSS project is to improve hospital bed design by applying features that would allow for increased efficiency, and reduce excess equipment needed by the nurse to apply to the bed and equipment that the patient would need. By having most of the equipment installed in the bed, it will lessen equipment that a patient requires during a hospital stay. This will also increase the number of devices accessible to all patients when needed, rather than looking for the items separately. With the added features, hospital staff will be able to provide more efficient and effective care for their patients.

Project Goals

The goal of the project is to improve the quality, efficiency, and usability of the hospital bed for patients, nurses, and doctors. The DFSS team will accomplish this by

1. Reviewing existing hospital bed products currently on the market and conducting our own market research
2. Developing seven different design concepts and evaluating them against the highest quality
3. Developing a model of the design concepts
4. Selecting and prototyping the chosen model

Requirements and Expectations

The requirements for this project entail collecting the voice of the customer (VOC), reviewing the current market literature regarding existing hospital beds, performing a design review, and analyzing hospital bed designs. The project team will develop a minimum of seven concept designs that meet or exceed current safety standards.

Project Boundaries

The project boundaries were established by

1. Identifying the customer
2. Defining customers' expectations and needs
3. Specifying deliverables linked to expectations
4. Identifying critical to quality (CTQ) for the deliverables
5. Mapping the process
6. Determining where in the process the CTQs can be most seriously affected
7. Evaluating which CTQs have the greatest opportunity for improvement
8. Defining the project to improve the CTQs selected

Designing hospital beds will require the team to collect information from the customer and meet certain design criteria in order to reduce entrapment and potential life-threatening entrapment areas within the hospital bed system. Following these boundaries will allow the team to create a rigid flowchart based on customer input and measurable attributes that would be most critical in the deliverables and within the guidelines of the Food and Drug Administration (FDA). The FDA regulates hospital bed activities by analyzing reports of product problems and adverse events. The design criteria will be regulated by Title 21, Code of Federal Regulations, Part 820—Quality System regulation and the International Electrotechnical Commission (IEC) 60601-2-38; Amendment 1, 1999 Medical Electrical Equipment—Part 2–38: *Particular Requirements for the Safety of Electrically Operated Hospital Beds.* Hospital bed systems are covered under the guidelines shown in Table 28.1.

Project Management

The team determined there were several aspects that were important for managing the project. The first aspect was stakeholder (social network) analysis, which involved

- Identifying stakeholders
- Gathering the required information
- Assessing stakeholders
- Developing a good understanding of the project expectations among group members
- Determining the main issues/concerns among stakeholders
- Prioritizing work among stakeholders
- Assign responsibilities among the team
- Communicating among stakeholders
- Informing team members about the project progress

TABLE 28.1

Hospital Bed System Guidelines

Product Code	CFR Section	Classification Name	Class
FNJ	880.5120	Manual adjustable hospital bed	I
FNK	880.5110	Hydraulic adjustable hospital bed	I
FNL	880.5100	ac-powered adjustable hospital bed	II
FPO	880.6910	Wheeled stretcher	II
IKZ	890.5225	Powered patient rotation bed	II
ILK	890.5150	Powered patient transport	II
INK	890.3690	Powered wheeled stretcher	II
INY	890.5180	Manual patient rotation bed	I
IOQ	890.5170	Powered flotation therapy bed	II

The team also identified several key areas to ensure project success including

- Proper planning
- Developing a rapport with key stakeholders
- Proper management of time and resources
- Learning how to identify and use the right tools
- Proper risk management and development of contingency plans
- Providing leadership without micromanaging
- Looking for warning signs

The team also developed a project plan for the project as shown in Figure 28.1. The project plan followed the DMADV roadmap as shown in Figure 28.2.

Invent/Innovate Phase

Voice of the Customer

The team began with identifying the customer requirements and the approach/system architecture that would best meet the project requirements. The Define phase involved developing a project charter, gathering the VOC, performing competitive analysis, and developing CTQs. Before establishing the project milestones, the team had to identify and cascade customer and product requirements. Documenting the VOC involved identifying the customers and their respective wants or needs through a survey. The customers were asked to answer a set of questions and were also asked to forward the survey to collect as much relevant information as possible. Subsequently, the team was able to collect 60 responses. The survey data was then analyzed to determine the critical characteristics with respect to the project objectives.

KJ Analysis and Kano Model

The main purpose of Kawakita Jiro (KJ) analysis is to integrate customer needs with the VOC to obtain a well-defined set of customer requirements. The team used brainstorming rather than creating affinity diagrams to gather and group the VOC wants and needs. Brainstorming helped the team to bring together various ideas and suggestions. The interesting proposals were refined to determine the critical information needed to shortlist various new, unique, and difficult needs. The team also used a Kano Model to gain a thorough understanding of the customer needs and generate inputs for quality function deployment (QFD). The Kano Model is shown in Figure 28.3. The

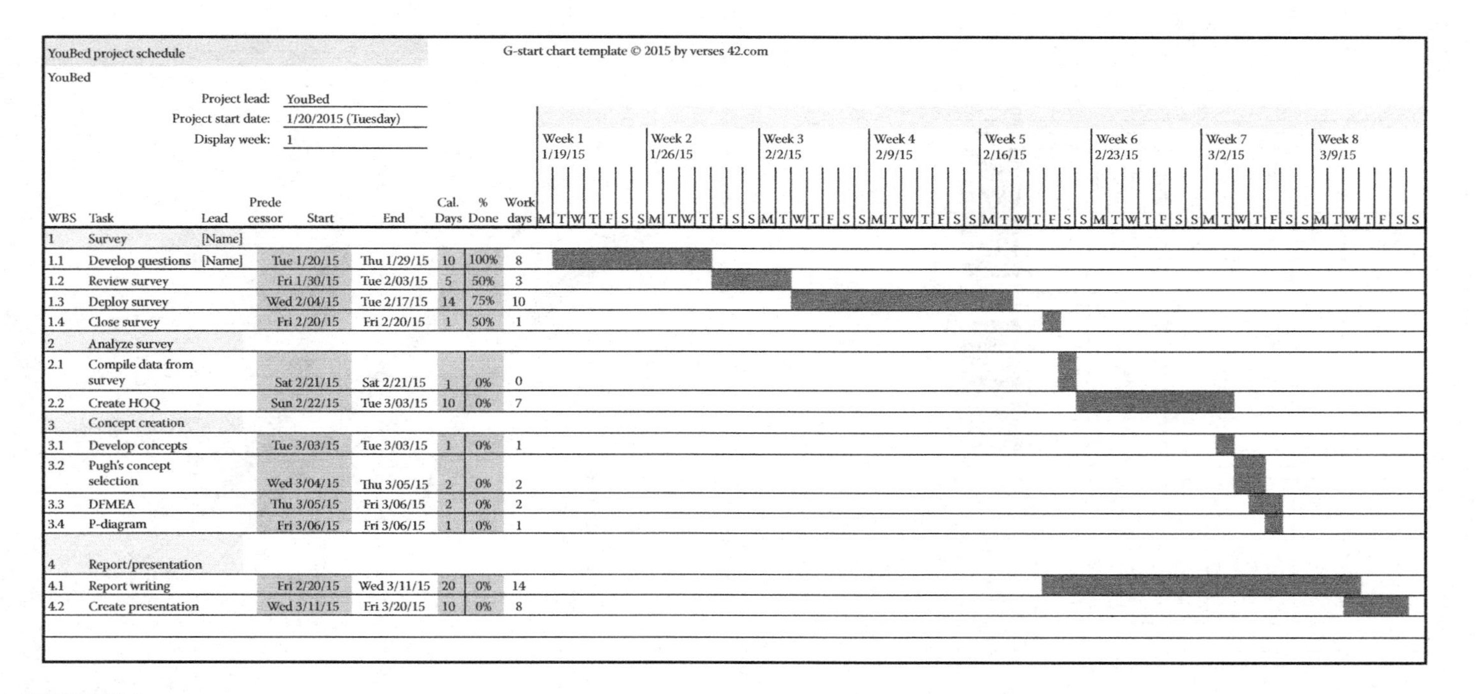

WBS	Task	Lead	Predecessor	Start	End	Cal. Days	% Done	Work days
1	Survey	[Name]						
1.1	Develop questions	[Name]		Tue 1/20/15	Thu 1/29/15	10	100%	8
1.2	Review survey			Fri 1/30/15	Tue 2/03/15	5	50%	3
1.3	Deploy survey			Wed 2/04/15	Tue 2/17/15	14	75%	10
1.4	Close survey			Fri 2/20/15	Fri 2/20/15	1	50%	1
2	Analyze survey							
2.1	Compile data from survey			Sat 2/21/15	Sat 2/21/15	1	0%	0
2.2	Create HOQ			Sun 2/22/15	Tue 3/03/15	10	0%	7
3	Concept creation							
3.1	Develop concepts			Tue 3/03/15	Tue 3/03/15	1	0%	1
3.2	Pugh's concept selection			Wed 3/04/15	Thu 3/05/15	2	0%	2
3.3	DFMEA			Thu 3/05/15	Fri 3/06/15	2	0%	2
3.4	P-diagram			Fri 3/06/15	Fri 3/06/15	1	0%	1
4	Report/presentation							
4.1	Report writing			Fri 2/20/15	Wed 3/11/15	20	0%	14
4.2	Create presentation			Wed 3/11/15	Fri 3/20/15	10	0%	8

FIGURE 28.1
DFSS project plan.

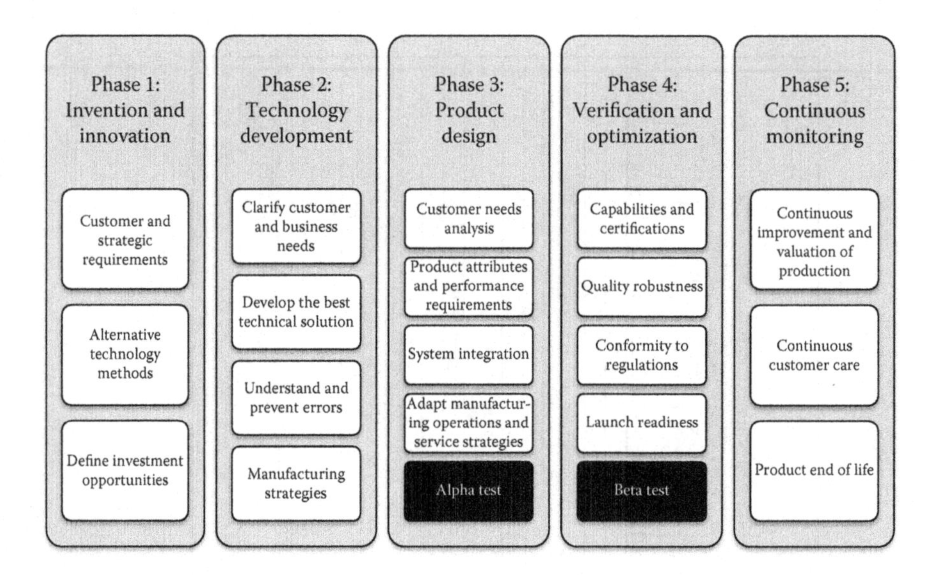

FIGURE 28.2
DFSS roadmap.

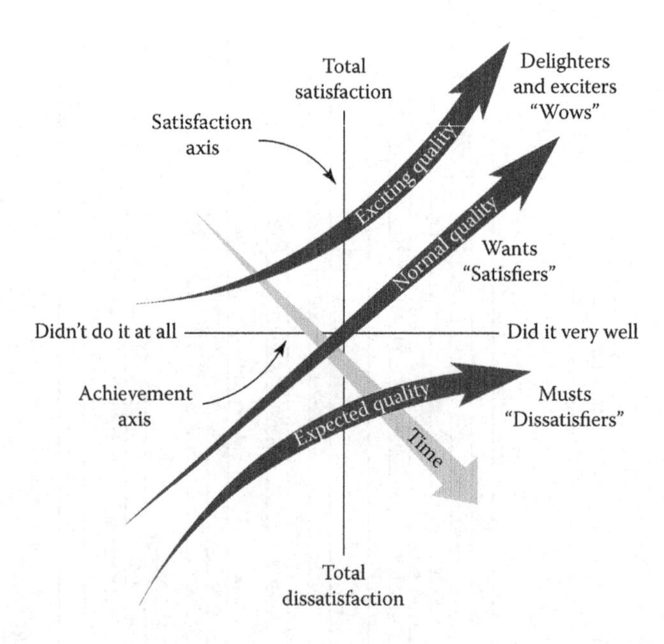

FIGURE 28.3
Kano Model.

TABLE 28.2

Kano Analysis of Customer Needs

Level of Customer Needs	Customer Needs
Exciting quality	• Built-in computer tablet for entertainment • Ability to track medications
Normal quality	• Adjustable height • Suitably placed controls • Right size • Ergonomic side rails • Easy to move around
Expected quality	• Comfortable • Affordable

customer needs for this DFSS project with respect to the level of customer needs based on the Kano Model are provided in Table 28.2

For this project, the team chose to focus on normal and expected qualities while ignoring the 'exciting qualities' as they added to the cost of the hospital bed, thus making it too expensive. After another round of brainstorming, the team grouped, ranked, and prioritized customer requirements. The customer requirements for the hospital bed were prioritized as follows:

1. Suitable height
2. Ergonomic controls
3. Right bed size
4. Comfort
5. Ergonomic side rails
6. Easy to move around in a room
7. Easy to move from room to room
8. Affordable price
9. Ability to track medications

The exciting, normal, and expected needs are translated to system-level technical requirements using QFD.

Quality Function Deployment

QFD is a systematic process to integrate customer requirements into different engineering targets to be met by the new product design. The team used the house of quality (HOQ) to document different customer requirements

TABLE 28.3

Customer and Functional Requirements

Demanded Quality (Customer Requirements)	Quality Characteristics (Functional Requirements)
Right bed size	Appropriate dimensions
Ability to track medications	Built-in equipment/adapter
Suitably placed controls	Ergonomically placed controls
Comfortable	Better mattress
Easy to move around in a room	Height-adjustable mechanism
Easy to move from room to room	Padded side rails
Ergonomic side rails	
Affordable price	

and respective functional requirements. Table 28.3 provides the respective customer and functional requirements.

The "customer requirements" or the "voice of customer" was gathered through the survey, which enabled the team to record the customers' needs and problems. The "quality characteristics" or "voice of company" were generated by the project team so as to meet the VOC.

The "interrelationships" were based on three relationships, strong, moderate, and weak, and helped the team to translate the VOC into the voice of the company/product.

For the "competitive analysis", the team compared the ability of three different beds to satisfy the customer requirements. The three different beds used to quantify the requirements were

1. Hill-Rom Advanta 2 Med Surg Hospital Bed
2. Invacare Carroll CS7 Hi-Low Hospital Bed Set
3. GoBed II Med/Surg Hospital Bed

The HOQ is shown in Figure 28.4.

Develop Phase

Design for X Methods and Concept Generation

Based on the VOC and the voice of the company, the team developed nine conceptual designs.

Conceptual Design 1: Scissor lift with adjustable height (shown in Figure 28.5)

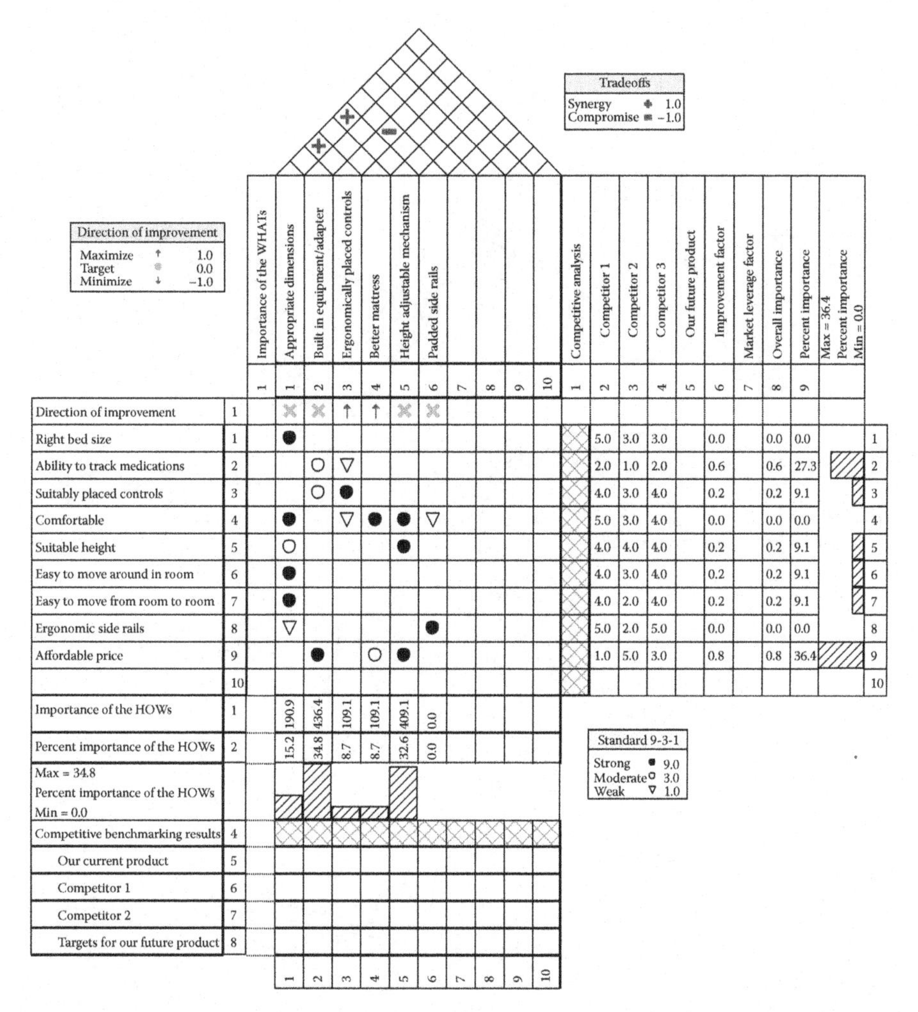

FIGURE 28.4
House of quality.

- This design uses a scissor lift to adjust the height of the bed.
- A motorized hydraulic unit can be used to adjust the height, which can be operated using a remote control integrated with the frame.
- A safety lock bar will be used to counter sudden and/or unwanted change in the height of the bed frame.
- Using a scissor lift mechanism will help achieve a small footprint, which also increases the ease of moving it around.

Conceptual Design 2: Cushioned armrest with built-in controls (shown in Figure 28.6)

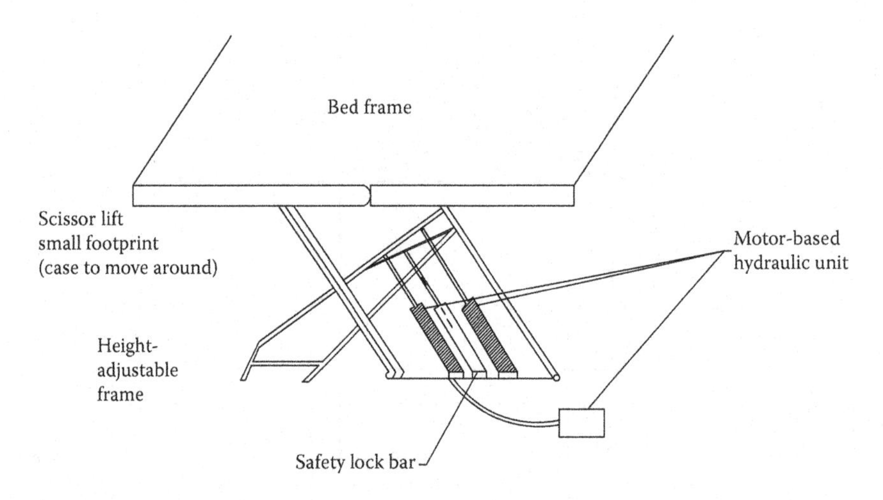

FIGURE 28.5
Conceptual Design 1: Scissor lift.

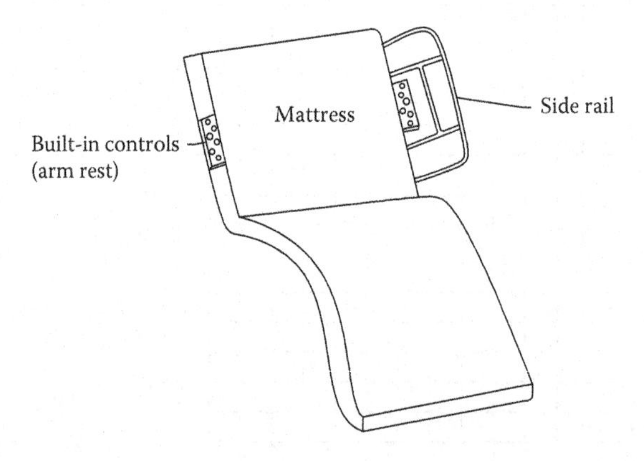

FIGURE 28.6
Conceptual Design 2: Cushioned armrest with built-in controls.

- A cushioned armrest with built-in (and ergonomically placed) controls can be used to make it easier for the patients to reach for the controls when needed.
- The armrest will also provide support for the arms and will be added to both sides of the bed.

Conceptual Design 3: Height-adjustable frame (shown in Figure 28.7)

- The height of the frame can be adjusted using a hydraulic/motor unit.
- This design will have a larger footprint than Concept Design 1 with the scissor lift mechanism to adjust the frame height.

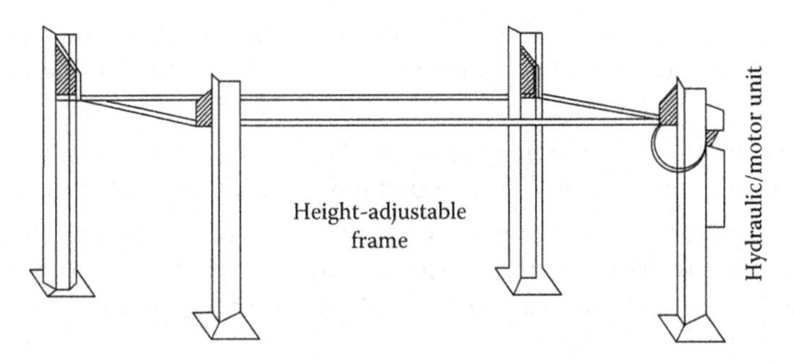

FIGURE 28.7
Conceptual Design 3: Height-adjustable frame.

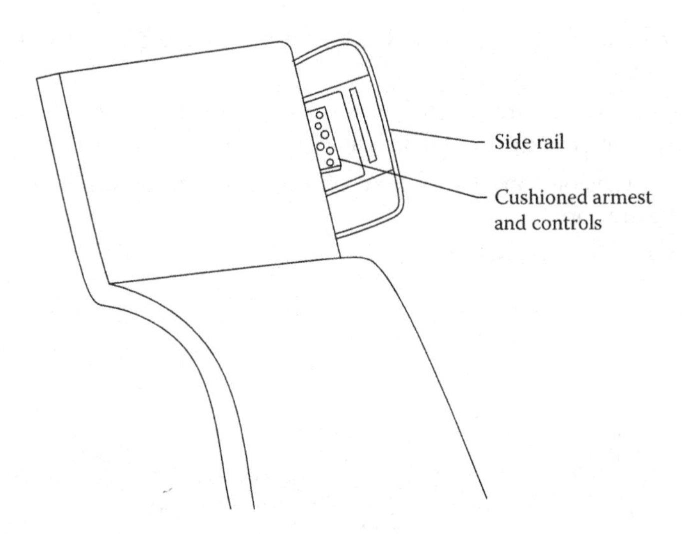

FIGURE 28.8
Conceptual Design 4: Padded side rails.

Conceptual Design 4: Padded side rails (shown in Figure 28.8)

- The padded side rails will help in preventing unnecessary/avoidable bruising of the patients.

Conceptual Design 5: Connection lines input/output, hydraulic A-frame with adjustable height (shown in Figure 28.9)

- This design uses an A-frame lift to adjust the height of the bed to a minimum of 6″.

- A motorized hydraulic unit can be used to adjust the height, which can be operated using a remote control integrated with the frame.
- Maximizing the distance that the bed can be lowered will allow staff to easily move patients from a bed to a chair or to another bed.
- Connection lines from source or monitors can be plugged directly in the bed safety rails, and then from the safety rails to the patient, which reduces the risk of patients getting tangled in their lines.
- This design minimizes the connection lines and equipment needed.

Conceptual Design 6: Extension bed (shown in Figure 28.10)

- An adjustable bed that will allow for extending the bed base to accommodate patient size.
- The extensions are within the makeup of the bed platform. This allows for the extension to be easily tucked away.
- The extension can be pulled out or slid in manually for easy configuration.
- Safety rails are attached to the extension arm.

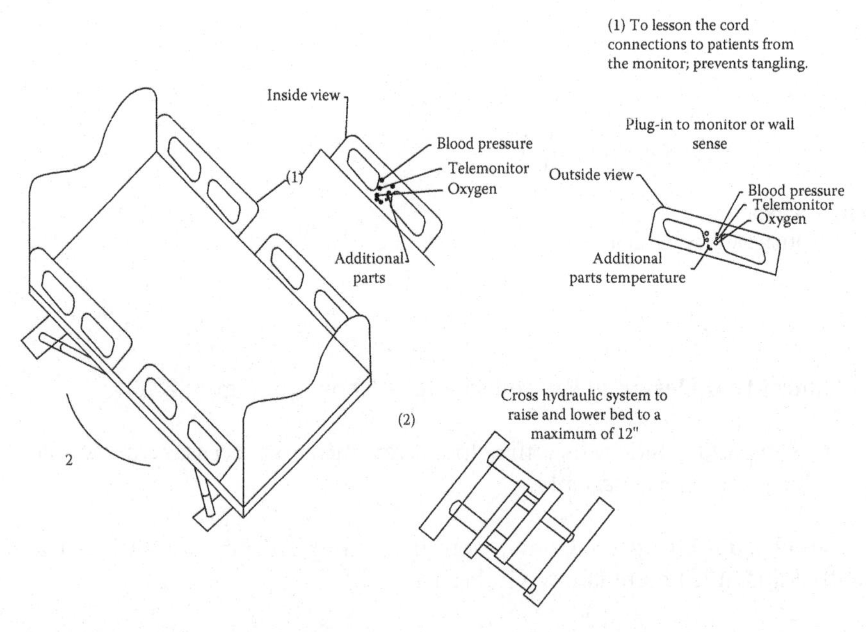

FIGURE 28.9
Conceptual Design 5: A-frame with hydraulic lift.

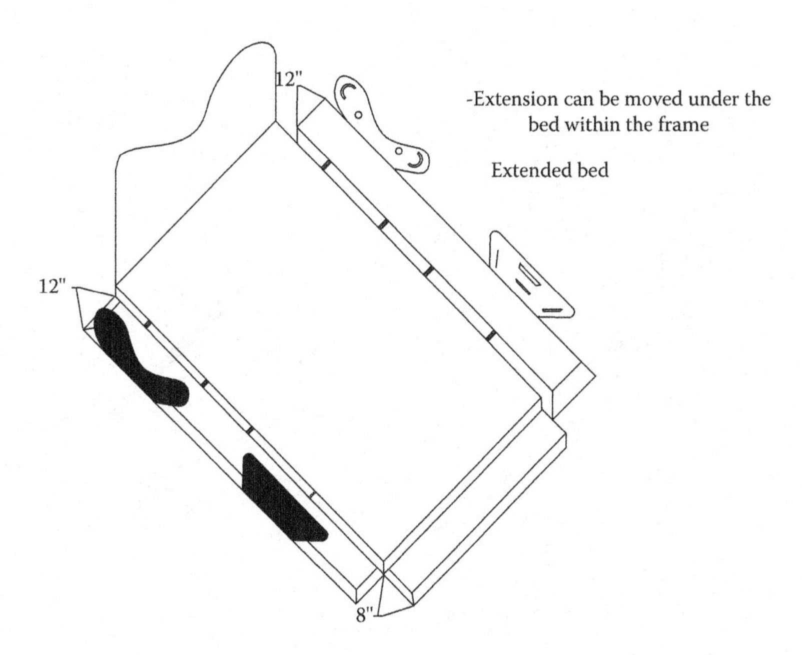

FIGURE 28.10
Conceptual Design 6: Extended bed.

Conceptual Design 7: H-frame adjustable height bed frame (shown in Figure 28.11)

- The height of the frame can be adjusted using a hydraulic/motor unit.
- This design will have more stability and can support more weight.
- The system is tucked away under the bed with safety guards on each side to avoid injury.
- Maximizing how low the bed can be set allows patients to be easily moved from one platform to another.
- Staff can easily move patients by electronically adjusting the height of the bed.

Conceptual Design 8: Patient and nurse storage (shown in Figure 28.12)

- Includes areas for storage of patients' miscellaneous items and nursing equipment.
- Side rails slide up and down to allow easy access in and out of the bed.
- Mattress is inflatable to allow the patient to control the comfort.
- Wireless remote that attaches to the side rails allows quick access for patient to adjust the bed to his or her needs.

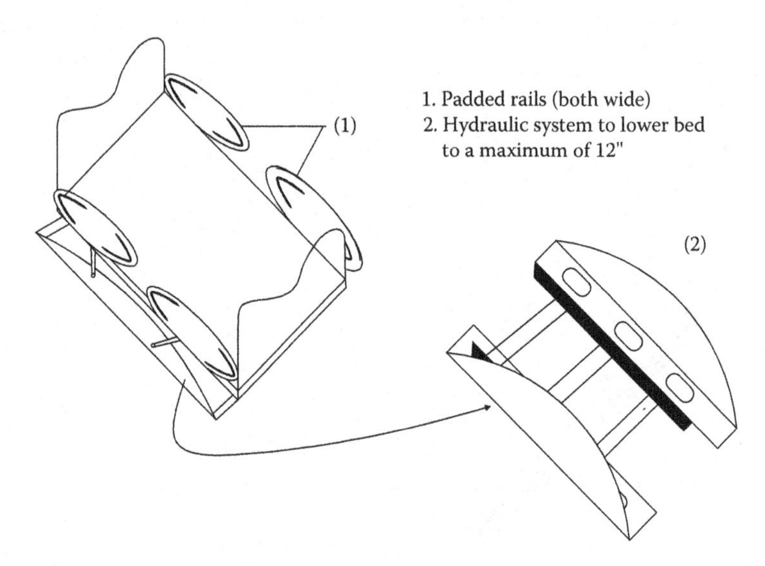

FIGURE 28.11
Conceptual Design 7: H-frame (adjustable).

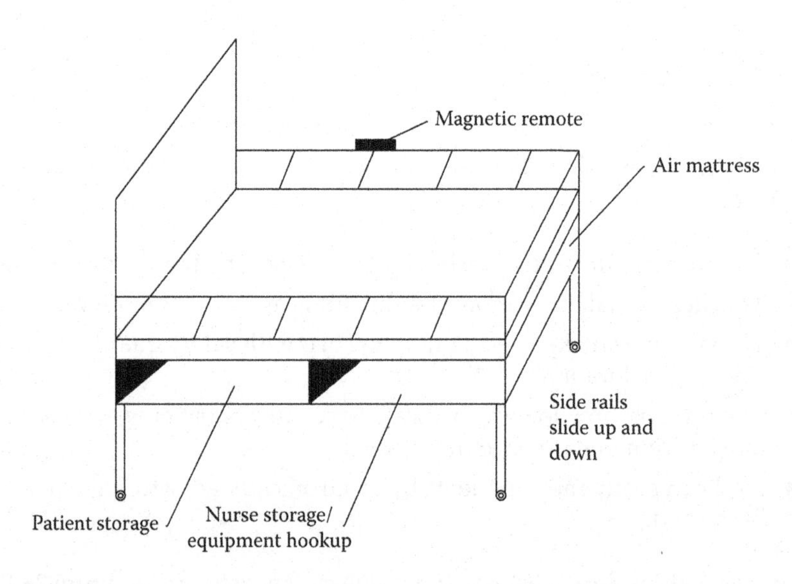

FIGURE 28.12
Conceptual Design 8: Patient and nurse storage.

Conceptual Design 9: Adjustable bed frame (shown in Figure 28.13)

- Bed frame can bend in three locations to optimize patient comfort.
- Mattress is inflatable to allow the patient to control the comfort.
- Mattress is heated and cooled for patient comfort.

- Side rails swing up to allow patients to easily get in and out of the bed.
- Shelf in front for nurse records.
- Pocket in side rails for patient items.

Table 28.4 shows the ranking of each of the concepts to the different criteria ratings for design and production. The scale uses ratings of 1–10; where, 1 = easiest to achieve and 10 = hardest to achieve.

Pugh Concept Selection Matrix

The Pugh Concept Selection Matrix (PCSM) was used to compare and select the best ideas/concepts using a weighted score, as shown in Table 28.5. It also helped to identify the best concept and create a hybrid solution, if needed. Using the PCSM, the team chose to focus on Concept 7 for the final prototype.

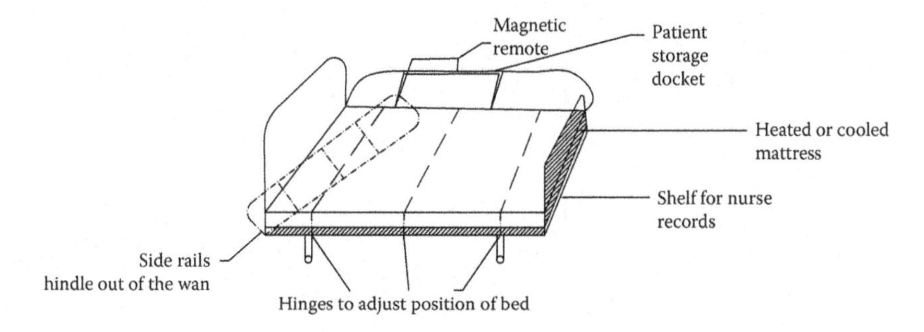

FIGURE 28.13
Conceptual Design 9: Adjustable bed frame.

TABLE 28.4
Design and Production Criteria Ratings

	Concepts								
	1	**2**	**3**	**4**	**5**	**6**	**7**	**8**	**9**
Producibility	7	5	6	5	9	5	7	7	7
Assemblability	6	6	8	5	8	3	5	5	5
Reliability	6	4	5	5	7	3	4	4	6
Serviceability	7	6	6	5	8	3	4	4	5
Environmentally friendly	6	6	6	6	5	3	5	7	6

TABLE 28.5

Pugh Concept Selection Matrix

			Concepts								
	Baseline	**Weight**	**1**	**2**	**3**	**4**	**5**	**6**	**7**	**8**	**9**
Suitable height	0	5	5	0	5	0	5	5	5	0	−5
Ergonomic Controls	0	1	0	1	0	1	1	0	−1	1	1
Easy to move	0	3	3	0	−3	0	0	−3	0	0	3
Comfortable	0	4	−4	4	−4	4	0	4	4	−4	4
Built-in medication tracking	0	1	0	0	0	0	1	0	0	0	0
Ergonomic controls	0	2	0	2	0	2	2	0	2	2	2
Affordable	0	1	0	1	1	1	0	1	1	1	0
SUM	—	—	4	8	−1	8	9	7	11	0	5

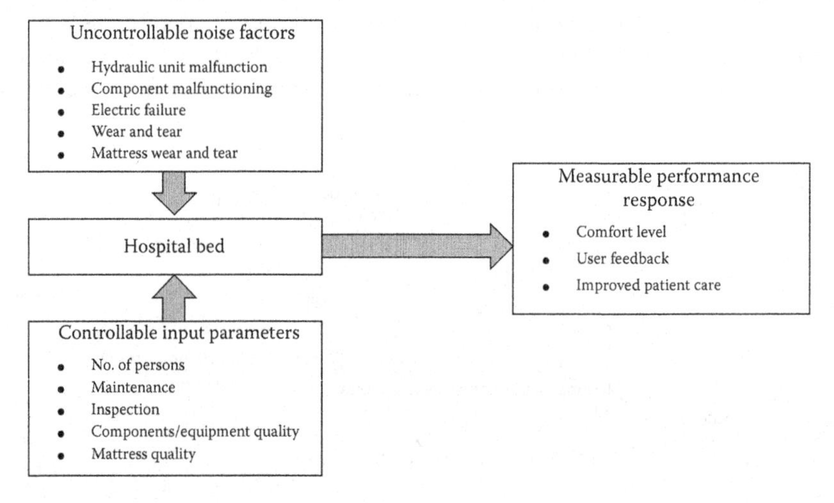

FIGURE 28.14
P-diagram.

Verify

The DFSS team used the P-diagram to evaluate the input–output relationship between the performance parameters (Figure 28.14). The feedback for P-diagram also aided in developing the design failure modes and effects analysis (DFMEA).

Optimize

DFMEA was utilized to optimize the current design. Table 28.6 provides the DFMEA. By utilizing this tool, the team was able to determine the

TABLE 28.6

Design Failure Modes and Effects Analysis

Design Function Requirements	Potential Failure Mode	Potential Effect(s) of Failure	Severity	Potential Cause(s)/ Mechanism(s) of Failure	Occurrence	Current Design Controls	Detection	RPN	Recommended Actions
Inflatable mattress	Pops or fails to inflate/ deflate	Injury to patient	8	Bed pump	2	Bed testing before shipment	4	64	
				Mattress punctured	1	Puncture proof fabric	1	8	
Hydraulic lift	Hydraulic fluid leak	Lift slow or inoperable	7	Design defect	2	Test before shipment	3	42	
	Remote sensor fail	Unable to operate	8	Damaged during shipment or install	2	Full bed test	5	80	
Wheels	Stuck wheel	Difficulty moving bed	4	Bent during operation	7	Strength test components	4	112	
				Casters not lubricated	7	Test before shipment	5	140	Include instructions to maintain caster lubrication
Side rails	Bent out of place	Patient unsafe	9	Misuse of rails or fatigue	3	Designed to withhold standard patient misuse	1	27	
Electronic controls	Not controlling bed	Patient stuck in same position	8	Manufacturing defect or abuse of equipment	4	Remote designed to be durable for drops up to 5 feet	4	128	

main aspects of the design that caused more risk than others, which included (1) wheels becoming stuck due to being bent during operation, (2) casters not being lubricated properly, and (3) electronic controls being defective due to manufacturing defect or abuse of equipment. With regard to wheels becoming stuck due to being bent or the electronics not working properly, the team determined that the current design controls monitor these problems sufficiently; therefore, no further actions were needed. For the casters not being lubricated, the team determined that is was necessary to include instructions on properly maintaining them and lubricating the wheels at a set frequency to ensure they are always in proper working order.

Tips to Improve Our Design

The team determined several additional steps that could be taken to further improve the design including

- Building a full working prototype
- Utilizing focus groups at hospitals of both nurses and patients to get input on the proposed design
- Working with the manufacturer to ensure the design can feasibly be built
- Conducting hospital trials to test the prototype
- Adding nurses to the design team to optimize the usability of the bed
- Stress testing all components to confirm they withstand standard product use

Final Design Prototype: Concept 7

A small-scale prototype was developed to test the proposed concept as shown in Figure 28.15:

- Legs with side safety guards
- Hydraulic housing unit
- Hydraulic cylinder
- Motor
- Power cord

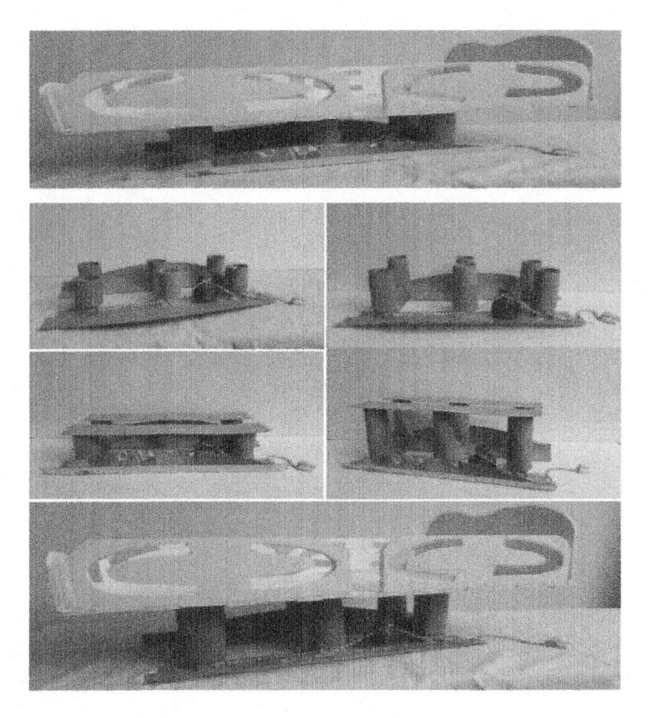

FIGURE 28.15
Prototype of the proposed concept.

Conclusion

The DFSS team believes that we have designed a bed concept that can revolutionize the hospital bed industry. With the DFSS principles applied to this project, this bed can be produced less expensively, be used by a wider range of clients, and last longer than current beds. The bed has been designed with both the patient and nurse in mind so that all involved with the bed can have the best possible experience.

Glossary

affinity diagram: a process to organize individual pieces of information by placing it on cards and grouping the cards that go together based on similarity. Header cards are then used to summarize each group of cards.

analysis of variance (ANOVA): a statistical procedure that uses variance to determine differences in means. ANOVA is used when you have discrete Xs with continuous Ys.

attribute data: data in the form of go/no-go or pass/fail (see discrete variable).

balanced scorecard: used to monitor and focus on factors critical for business success, presented using four perspectives (competencies, business processes, customers, and financial).

benchmarking: an activity to establish internal expectations for excellence, based on direct comparison with the "best." In some cases, the best is not a direct competitor in your industry.

Black Belt: a person trained to execute critical projects for breakthrough improvements to enhance the bottom line. An individual within an organization who is an expert in applying Six Sigma tools and methodologies.

brainstorming: a technique to generate a large number of ideas in a short period of time and to build consensus among team members. The technique requires people to think of possibilities without evaluation.

breakthrough planning: a method for linking upper-level strategies to projects and aligning projects and resources.

capability: a measure, both graphical and numerical, of the ability of a process or product to meet a customer's requirements. Capability is typically expressed in Z, DPMO, or C_{pk}.

causality: the assertion that a change to an input factor will result in a change to an output.

cause and effect diagram: a problem-solving tool used to establish relationships between effects and multiple causes. It is a pictorial diagram in the shape of a fishbone showing possible causes (process inputs) for a given effect (process output). Also known as an Ishikawa diagram or fishbone diagram.

central tendency: a measure of the point about which a group of values is clustered. Examples of measures of central tendency include mean and median.

champion: an individual who acts as the sponsor or owner of a project and has the authority and responsibility to inform, support, and direct a team. The champion is responsible for communications and for breaking down barriers to make sure projects keep moving.

Typically, the individual is a director or executive-level manager. Also known as a mentor or sponsor.

common causes of variation: sources of variability in a process that are truly random and inherent in the process itself.

concurrent engineering: the practice of designing a product (or service), its production process, and its delivery mechanism simultaneously. The process requires considerable up-front planning as well as the dedication of resources early in the development cycle. The payoff comes in the form of shorter development time from concept to market, higher product quality, lower overall development cost, and lower product or service cost.

continuous response: a response or output that can be take any value within a set of limits.

control chart: a statistical tool for problem solving that indicates control of a process within established limits. A control chart consists of a run chart with statistically determined upper and lower control limits and a center line.

control element: a specific process variable that must be controlled. Measurements of a control element indicate whether or not a stable condition has been achieved.

control limits: upper and lower bounds in a control chart that are determined by the process itself. Control limits can be used to detect special causes of variation. Typically, control limits are set at ±3 standard deviations for the center line.

correlation coefficient: a measure of the linear relationship between two variables.

cost of poor quality (COPQ): the business cost that is incurred as a result of internal and external failures in a product or process. Costs associated with not doing things right the first time. Examples of COPQ include scrap, rework, and waste.

countermeasures: immediate actions taken to bring performance that is tracking below expectations into the proper trend. Requires root cause analysis.

C_p: a process capability measure defined as the ratio of specification width (USL–LSL) to process width (6σ). C_p is a capability index that shows the process capability potential but does not consider how the process is centered. Larger values indicate greater potential capability.

C_{pk}: a capability index that is used to compare the natural tolerance of a process with the specification limits. C_{pk} is equal to C_p if the process is centered on nominal.

critical factor: a CTQ that should not fail once it is designed into a product or process.

critical to quality (CTQ): an attribute or parameter of a part, service, or process that is directly related to the wants and needs of the customer. It should view quality from a customer standpoint. CTQs have two characteristics: they are important to the customer and measurable.

CTQ flowdown: a qualitative technique for linking high-level customer CTQs to lower-level design CTQs and *X*s.

customer: anyone who uses or consumes a product or service. A customer can be internal or external to the provider: the person or organization that will be the benefactor of the product or service. In Six Sigma, there are often both internal and external customers.

defect: a nonconformance in a product or service. Any event that is outside the customer's specifications.

defects per million opportunities (DPMO): a measure of process performance based on the number of defects. Each product or process can have multiple opportunities for a defect. DPMO is calculated by dividing the number of defects by the total number of opportunities and then multiplying this value by one million.

define, identify, design, optimize, verify, monitor (DIDOVM): the phases of the proactive Six Sigma methodology used for new product/process development.

define, measure, analyze, design, verify (DMADV): the phases that some companies use for their DFSS methodology.

define, measure, analyze, improve, and control (DMAIC): the phases of the Six Sigma methodology used for existing product/process improvement.

design dashboard: a spreadsheet that summarizes the CTQs for a new product or process that is under development.

design for manufacture and assembly (DFMA): a philosophy that strives to improve costs and employee safety by simplifying the manufacturing and assembly process through product design.

Design for Six Sigma (DFSS): a data-driven approach to reducing variation in the design of products and services to meet or exceed customer expectations and achieve 3.4 defects per million opportunities. The methodology for developing new products and processes that deliver Six Sigma performance on rollout. We use IDOV as the DFSS methodology. DFSS and IDOV are used interchangeably.

design of experiments (DOE): a structured approach to maximize gain in product or process knowledge. An experimental technique that enables experimental factors to be changed simultaneously without losing the ability to separate the effects of each factor.

deviation: the difference between an observed value and the mean of all observed values.

discrete variable: a random variable that can only result in two possible outcomes. For example, pass/fail in a functional test.

experimental design: purposeful changes to inputs or factors of a process to observe the corresponding changes in the outputs or responses.

factor: an input to a process that can be used during experimentation. These variables (Xs) are varied during the course of an experiment.

failure mode and effects analysis (FMEA): a technique used to manage risk. A structured approach to assess the magnitude of potential failures and identify the sources of each potential failure. Corrective actions are then identified and implemented to prevent failure occurrence.

fault tree analysis (FTA): a technique to evaluate the possible causes that may lead to product failure.

fishbone diagram: see cause and effect diagram.

five whys: a simple problem-solving method of analyzing a problem or issue by asking "Why" five times. The root cause should become evident by continuing to ask why a situation exists.

flowchart: a pictorial representation of a process that illustrates the inputs, main steps, branches, and outcomes of a process. A problem-solving tool that illustrates a process. It can show the "as is" process or the "should be" process for comparison and should make waste evident.

fractional factorial DOE: a DOE in which a limited number of factor combinations are tested. Fractional factorial DOEs lose some information but save time and cost.

full factorial DOE: a DOE in which all possible combinations of the factors are tested. Full factorial DOEs retain all possible information for evaluation.

gauge capability study: a method of collecting data to assess the variation in the measurement system and compare it with the total process variation.

gauge repeatability and reproducibility (R&R): a measure of a gauge's ability to provide the same results multiple times using different operators and tools.

Green Belt: an individual trained to assist a Black Belt. This individual may also undertake projects of a lesser scope than Black Belt projects.

histogram: a chart that displays data in a distribution, generally in graph form. It depicts the frequencies of numerical or measurement data. It may be used to reveal the variation that any process contains.

input: a resource that is consumed, used, or added during a process.

input–process–output (IPO) diagram: a visual representation of the inputs of a process, the process that transforms the inputs, and the resulting outputs.

Ishikawa diagram: see cause and effect diagram.

just-in-time (JIT): a strategy that concentrates on delivering the right products in the right time at the right place. This strategy exposes waste and makes continuous improvement possible.

key performance indicator (KPI): a method for tracking or monitoring the progress of existing daily management systems.

lower control limit (LCL): the limit above which the process subgroup statistics must remain for the process to be in control. Typically, the LCL is three standard deviations below the center line.

lower specification limit (LSL): the lowest value of a product specification for the product to be acceptable to the customer.

main effect: a measure of the effect of an individual factor removing the effect of all other factors.

Master Black Belt (MBB): an individual who trains and mentors others in Six Sigma tools and methodologies.

mean: the arithmetic average of a set of values. It is calculated by adding the sample or population values together and dividing by the number of elements (n). $\bar{x}$ denotes a sample mean; μ denotes a population mean.

mean time between failures (MTBF): the mean time between successive failures of a repairable product. MTBF is a measure of product reliability.

measurement system: a system that is used to measure a CTQ. All measurement systems should have an acceptable gauge R&R.

measurement system analysis (MSA): used to determine the reliability of a measurement instrument or gauge.

median: the middle value of a data set when the values are arranged in ascending or descending order.

metric: a performance measure that is linked to the goals and objectives of an organization.

multigenerational plan (MGP): a planned introduction phasing for product and process launch. MGP allows the team to phase in more risky features and functionality.

noise: unexplained variability in a response.

non-value-added (NVA): those process steps that take time, resources, or space, but do not transform or shape the product or service toward that which is sold to a customer. These are activities that the customer would not be willing to pay for.

opportunity: any of the total number of chances for a defect to occur.

output: a product or service delivered by a process.

Pareto chart: a vertical bar graph for attribute or categorical data that shows the bars in descending order of significance, ordered from left to right. Helps to focus on the vital few problems rather than the trivial many. An extension of the Pareto principle suggests that the significant items in a given group normally constitute a relatively small portion of the items in the total group. Conversely, a majority of the items will be of relatively minor significance (i.e., the 80/20 rule).

PDCA cycle: plan-do-check-act cycle. PDCA is a repeatable four-phase implementation strategy for process improvement. Sometimes referred to as the Deming or Deming cycle.

poka yoke: a Japanese expression meaning "mistake proof." A method of designing production or administrative processes that will, by their nature, prevent errors. This may involve designing fixtures that will not accept an improperly loaded part.

process: an activity that blends inputs to produce a product, provide a service, or perform a task.

process capability: a comparison of the actual process performance with process specification limits. Measures of process capability include, but are not limited to, C_p, C_{pk}, dpm, and σ_{level}.

process control: a process is said to be in control if all special causes of variation have been removed and only common cause or natural variation remains.

process map: a visual representation of the sequential flow of a process. Used as a tool in problem solving, this technique makes opportunities for improvement apparent.

pull: a system in which replenishment does not occur until a signal is received from a downstream customer.

push: conventional production in which product is pushed through operations based on sales projections or material availability.

quality characteristic: an aspect of a product that is vital to its ability to perform its intended function.

quality function deployment (QFD): a system for translating customer requirements into appropriate company requirements at each stage from research and product development, to engineering and manufacturing, to marketing/sales and distribution. Makes use of the voice of the customer throughout the process.

range: a measure of variability in a data set. The range is the difference between the largest and the smallest value in a data set.

reliability: the probability that a product will function properly for a specified period of time, typically under specified conditions.

return on investment (ROI): a profitability ratio that represents the benefit from an investment. It is calculated by dividing the net profit by the total assets.

rework: an activity to correct defects produced by a process.

robust design: a term generically used to describe the ability to make a product or process design insensitive to sources of variation.

root cause: the ultimate reason for an event or condition.

run chart: a graphical tool that illustrates a process over time.

scatter diagram: a chart in which one variable is plotted against another to determine whether there is correlation between the two variables.

scatterplot: a two-dimensional plot that displays bivariate data. See scatter diagram.

screening design: an experiment designed to separate the significant factors from the insignificant factors.

sigma (σ): sigma has two meanings in the context of Six Sigma. Sigma is the measure of quality, as in Six Sigma. Sigma is also the Greek symbol that is used to describe the standard deviation of a statistical population.

sigma capability: a measure of process capability that represents the number of standard deviations between the center of a process and the closest specification limit. See sigma level.

sigma level: a measure of process capability that represents the number of standard deviations between the center of a process and the closest specification limit. See sigma capability.

Six Sigma: a quality improvement and business strategy that emphasizes impacting the bottom line by reducing defects, reducing cycle time, and reducing costs. Six Sigma began in the 1980s at Motorola. An all-inclusive methodology (reactive and proactive) for selecting and executing projects focused on identifying and satisfying customer needs.

special causes of variation: nonrandom causes of variation. Control charts can be used to detect special causes of variation.

specification limits: customer-driven boundaries of acceptable values for a product or process.

stability: a process is said to be stable if there are no recognizable pattern changes and no special causes of variation are present.

stakeholder: a person who will be impacted by the product or process when completed.

standard: a prescribed, documented method or process that is sustainable, repeatable, and predictable.

standard deviation: a measure of variability in a data set. It is the square root of the variance.

standardization: the system of documenting and updating procedures to make sure that everyone knows clearly and simply what is expected of them. Essential for the application of the PDCA cycle.

statistical control: a process is said to be in statistical control when it exhibits only random variation.

statistical process control (SPC): the application of statistical methods in the control of processes. Generally, the emphasis with SPC is on tools, and specifically on SPC charts that plot performance over time, comparing the performance with control limits. These graphical and statistical methods are used for measuring, analyzing, and controlling variation in a process for continuous improvement.

total revenue: the price of a product multiplied by the quantity sold in a given time period.

upper control limit (UCL): the upper limit below which a process must remain to be in control. Typically, the UCL is three standard deviations above the center line.

upper specification limit (USL): the highest value at which a product is acceptable to the customer.
user: a person who will use or operate the product or process.
utility: the pleasure or satisfaction obtained from a good or service.
value: a capability provided to a customer for an appropriate price.
value added: any process or operation that shapes or transforms the product or service into a final form that the customer will purchase.
variability: a process is said to exhibit variation or variability if there are changes or differences in the process.
variance: a measure of variability in a data set or population. Variance is equal to the squared value of the standard deviation.
variation: see variability.
voice of the customer (VOC): desires and requirements of the customer at all levels, translated into real terms for consideration in the development of new products, services, and daily business conduct.
waste: also known as *muda*. Any process or operation that adds cost or time and does not add value. Eight types of waste have been identified:

1. Waste from overproduction.
2. Waste from waiting or idle time.
3. Waste from unnecessary transportation.
4. Waste from inefficient processes.
5. Waste from unnecessary stock on hand.
6. Waste of motion and efforts.
7. Waste from producing defective goods.
8. Waste from unused creativity.

Xs: the most critical elements to control to keep the *Y* under control. Similar to root cause.
Ys: the main response being measured in a project. *Y*s should be a CTQ.
Z_{LT} (sigma long term): a measure of the process capability over a time period in which all sources of variation are included in the data. Z_{LT} is calculated by determining the number of standard deviations that fit between the mean and the specification limits of the long-term data.
Z_{ST} (sigma short term): a measure of the process capability over a time period in which not all the sources of variation are included in the data. Z_{ST} is calculated by determining the number of standard deviations that fit between the mean and the specification limits of the short-term data.

Index